# Radioactive Waste
# Disposal

# Pergamon Titles of Related Interest

**Carter, et al.** MANAGEMENT OF LOW-LEVEL RADIOACTIVE WASTE
**Colglazier** POLITICS OF NUCLEAR WASTE
**El-Hinnawi** NUCLEAR ENERGY AND THE ENVIRONMENT
**Hall** RADIATION AND LIFE
**Jackson** NUCLEAR WASTE MANAGEMENT: The Ocean Alternative
**Urquhart** ATOMIC WASTE AND THE ENVIRONMENT

# Related Journals*

THE ENVIRONMENTAL PROFESSIONAL
MATERIALS AND SOCIETY
MATERIALS RESEARCH BULLETIN
NUCLEAR AND CHEMICAL WASTE MANAGEMENT
PROGRESS IN NUCLEAR ENERGY

**\*Free specimen copies available upon request.**

Volume 1: The Waste Package

# Radioactive Waste Disposal

**Rustum Roy**

*The Pennsylvania State University*

**Pergamon Press**

*New York   Oxford   Toronto   Sydney   Paris   Frankfurt*

621.483
R 888

Pergamon Press Offices:

**U.S.A.**  Pergamon Press Inc., Maxwell House, Fairview Park, Elmsford, New York 10523, U.S.A.

**U.K.**  Pergamon Press Ltd., Headington Hill Hall, Oxford OX3 0BW, England

**CANADA**  Pergamon Press Canada Ltd., Suite 104, 150 Consumers Road, Willowdale, Ontario M2J 1P9, Canada

**AUSTRALIA**  Pergamon Press (Aust.) Pty. Ltd., P.O. Box 544, Potts Point, NSW 2011, Australia

**FRANCE**  Pergamon Press SARL, 24 rue des Ecoles, 75240 Paris, Cedex 05, France

**FEDERAL REPUBLIC OF GERMANY**  Pergamon Press GmbH, Hammerweg 6 6242 Kronberg/Taunus, Federal Republic of Germany

**Library of Congress Cataloging in Publication Data**

Roy, Rustum.
  Radioactive waste disposal.

  Includes bibliographical references.
  Contents: v. 1. The waste package.
  1. Radioactive waste disposal.  I. Title.
TD898.R66     1982      621.48'38      82-393
ISBN 0-08-027541-9 (v. 1)                AACR2

**Printed in the United States of America**

# PREFACE TO THE SERIES

This series of monographs on the science underlying the future technology of radioactive waste disposal has two purposes. The first purpose is to provide in a particular volume an authoritative review of one of the major components of the RWM-system. Each individual review is specifically designed to provide a balanced, well-proportioned, treatment of the subject. The aim is not primarily to be either up to the minute or exhaustive, but to be clear and authoritative. Second, it is designed to set any scientific knowledge, new research or technological advance in *perspective*. What difference will any particular work make? How has it changed our knowledge of or prospects for the whole system? This leads us to repeat in each monograph the introductory chapter on "the system" and the importance of always keeping the system in view.

The readership to whom this series is addressed is the entire science community interested in RWM. The geologist or nuclear physicist interested in the waste form, or the metallurgist interested in farfield migration or the seabed as repository should find in these monographs an authoritative introduction of the subject and a review of the major developments in the field up to approximately a year before the publication date.

The authors of each of the monographs has therefore been chosen from among those who have had considerable experience with their own specialty related to RWM, but also have a catholic view of the overall problem. By choosing most authors from within the relatively small academic community concerned with the problem we have attempted to avoid the advocacy of particular national agency approaches. The normal scientific biases remain, of course!

# INTRODUCTION TO THE SERIES

## Nuclear Waste: Origins, Amounts, Locations

The harnessing of nuclear fission as a source of power was a very significant technological innovation. Although the early hopes for abundant and cheap energy have fallen far short of everybody's predictions since the middle 1950's, the fact of the existence of a substantial and slowly growing nuclear power industry brings in its wake a rather special problem. Nuclear wastes were ignored by the technological designers of the system for two decades. Yet they are unique as waste products in many aspects; and the association with radiation and hence cancer has caused this "back end of the fuel cycle" to assume an importance far in excess of its technological significance.

The term nuclear wastes, in fact, encompasses four very different classes of material and it is important to distinguish these one from the other.

Figure I shows schematically how the different kinds of nuclear waste originate from different parts of the fuel cycle.

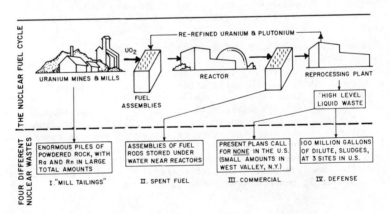

Fig. I. *The four major, distinctly different categories of nuclear waste and where they originate from within the nuclear fuel cycle are illustrated in this drawing. In most countries, "spent fuel" is not called a waste; in the U.S., it may one day be treated as such. Note that defense and commercial high-level wastes are very different materials. Not included in this drawing are low-level and medical nuclear wastes.*

*Mill tailings*, the sludges left from extracting uranium from crushed rock, exist in hundreds of millions of tons spread over dozens of locations in many western states such as Colorado, Utah, and New Mexico, and at similar sites all over the world.

*Spent* (i.e., used) *fuel* from power-generating reactors is stored at some 60 reactor sites in the U.S. and others in a dozen other countries, awaiting national policy decisions on its reprocessing. Spent fuel consists mainly (96%) of rather stable uranium oxide, with 1% of $PuO_2$ and 3% of fission products encased in zircaloy tubes.

*Commercial waste* is the material that is derived from the dissolution of the power reactors' spent fuel rods (after chopping up) in acid, and the separation by complexing and precipitation of the uranium and plutonium compounds. Those are returned as fuel, and the acid liquid stream becomes high-level liquid wastes (HLLW). There is virtually no *commercial waste* from reprocessing of power reactor fuel in the U.S. at present because, except for a small amount stored at West Valley, New York, no such reprocessing has ever been done. In Britain, the USSR, Japan and France, such reprocessing is just starting on a big scale.

On the other hand, reprocessing of fuel to make nuclear weapons, has gone on steadily for nearly 40 years. This has left the U.S. with 100 million gallons of high-level *"defense" wastes*; and the other nuclear weapons states (USSR, Britain, France and China) have correspondingly large amounts. *One of the major errors of nontechnical policy makers is to equate (U.S.) defense wastes with commercial wastes.* In fact, U.S. defense wastes are very different, being 100 times more dilute, chemically basic, and consisting of complex "dirty" sludges composed mainly of Al-Fe-oxyhydroxides, in contrast to the clear, acidic liquids of reprocessed commercial waste.

The existence of these differences among the four kinds of waste is hardly realized by most scientists and engineers. Nearly 99 percent of the public's attention is focussed on commercial wastes that do not even exist yet in many countries. And as is often the case with esoteric technologies, the greater dangers to the public probably exist in the least publicized areas, in this case, mill tailings. From millions of tons of these, which are lying on open ground, radon-222 gas and radium-226 (much more toxic than plutonium) are being extracted by air and water. Although hardly anything to panic about, mill tailings are much more difficult to manage than high-level waste. The technologically significant fact about all the other nuclear wastes is that there is so little of them, and they are geographically concentrated and well guarded.

# THE RADIOACTIVE WASTE DISPOSAL "SYSTEM"

The most important single concept regarding radioactive waste disposal which is grossly neglected by most scientists and laypersons alike, is the nature of the "SYSTEM" which it constitutes. The management of radioactive waste is *not* a technological issue alone: it is equally a socio-political issue and simultaneously a complex issue in the legal and regulatory scene involving State, Federal and international levels. Figure II is a schematic way of showing the interactions between these three major subsystems: technical, socio-political, and regulatory. Changes in any one affect the other two. A new regulation proposed by the U.S. Nuclear Regulatory Commission for instance, will have profound impacts on the technological subsystem. For example, if the proposed rule regarding a fifty year period of waiting after emplacement before final closure of the repository is enacted it might well rule out salt as a host rock. Technological advances in waste form insolubility or durability of concrete structures may slowly build up public confidence that more than one repository rock type may be available for use in different states. Conversely public education over a decade may substantially change the *perception of risk* bringing it closer to other comparable risks. This may permit very different technologies to be deployed which today are excluded mainly because of the public's anxiety. A very significant example of this is the use of the mid-gyre seabed sediments as the host rock for containing wates. The international public's attitudes toward *any* use of the sea has been shaped by the reactions to surface ocean "dumping" of low level wastes. If a worldwide scientific consensus emerges that the careful emplacement under the seabed of highly insoluble waste forms is among the safest and most cost-effective solutions, this could become part of a worldwide education effort on the real hazards from nuclear wastes and other sources. Such an education campaign resulting in an international agreement on seabed emplacement with a payment schedule into a United Nations development fund, could totally change the main technologies for the waste form, waste package and emplacement. Yet while we stress the awareness of all these subsystems of Fig. II in the total RWM system, *this* series of monographs focusses principally on the technical subsystem.

The first observation on the technical subsystem for radwaste disposal is that it in turn is also a system consisting of many interacting components. The system-elements are shown schematically in Fig. III, and they have exactly the same interactive characteristics among themselves as the larger subsystem.

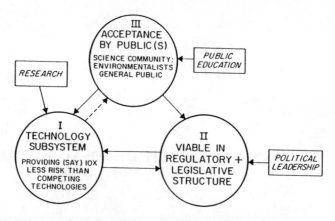

Fig. II. The total SYSTEM for Radioactive Waste Management.


Each of the elements: Storage, the waste solidification into the waste package, the physical isolation are separate means by which the hazardous radionuclides, $Cs^{137}$, $Sr^{90}$, $Tc^{99}$, and the actinides are prevented from reaching the biosphere and hence, humanity.

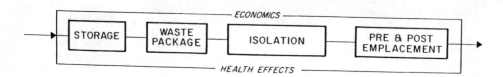

Fig. III. The components of the technology subsystem.


Storage: Refers to the storage chiefly of spent fuel, typically underwater in pools next to the reactors. However, plans for storing spent fuel in casks in large above-ground storage bunkers are well along in many countries. Radioactive decay diminishes the threat, and the magnitude of this change is very significant in reasonable times (the heat generated drops by nearly a factor of 10, in a few years as compared to the out of reactor condition). Hence how and for how long one stores fuel becomes very significant in the total strategy.

Waste package: The waste package refers to the sequential assemblage of solids which by physical barriers or chemical interaction, immobilize the radionuclides in repository condition-resistant materials. While a radionuclide is so "tied up" on an atomic scale inside a *solid* phase it is obviously not available to be transported, since solid state diffusion rates are so slow.

Isolation: The principal strategy of the waste disposal community for the last thirty years was to physically isolate the wastes from the biosphere. The favorite method was to place the wastes in a mined cavity ∿500 meters deep in the earth. The radionuclides erroneously assumed to be totally dissolved, were then shown not to be able to reach the biosphere. Note that the only feasible means for transport has to be via a *liquid* phase.

Emplacement: Deals with the process of emplacing the wastes--the transportation and the movement into the mines, and the sealing up of all boreholes and shafts, to prevent either solids or liquids from getting in or out of the repository.

It can easily be imagined how these four elements interact so that choices in any one affects profoundly the choices in the others. This has been discussed in detail by the author (1), and its consequences for waste management policy were first spelled out by the National Academy of Sciences, Committee on Radioactive Waste Management, in its Panel Report on Waste Solidification (2). For example, increasing the length of "storage" time profoundly affects both the waste form design (because of the radio-nuclide content and surface temperature at the waste form interface) and the repository design (due to the decreased heat loading). Likewise, choosing a particular host rock will certainly change the *optimum* waste form, or canister. In Fig. III, there are listed some of the major elements within each of the components of the technical subsystem. Each of these columns in Fig. IV is, of course, a subsystem of redundant barriers--a sort of subsystem within a subsystem within a subsystem. The image of the Russian Matryoshka dolls, nestled one inside the other, suggests itself as it did to us in 1973 as a visual description of this kind of relation.

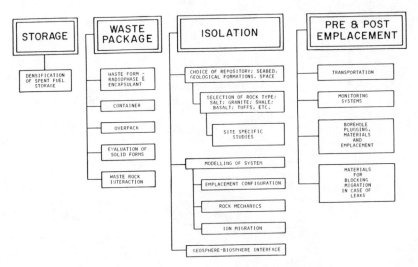

*Fig. IV. The research areas within each of the technology subsystem components.*

(1) National Research Council, Committee on Radioactive Waste Management, Panel on Waste Solidification (1978), Solidification of High Level Radioactive Wastes, National Academy of Sciences, Washington, DC.

(2) R. Roy (1979), Science Underlying Radioactive Waste Management: Status and Needs, Scientific Basis for Nuclear Waste Management, Vol. 1, Ed. G.J. McCarthy, Plenum Press, NY, p. 1.

The purpose of this detailed elaboration on the reality of this multi-barrier system is to indicate how difficult it is for really valuable science to be done on any one aspect of radioactive waste disposal without an awareness of all the connections to other parts of the system. For the purposes of the reader of a volume in this series, the figures presented in this introduction provide a framework for linking any aspect of current research to the total system for nuclear waste disposal. This *awareness of the whole when examining a particular part of the system* provides the proper perspective for the reader to assess the significance of any changes effected by research, legislation, or public attitudes.

Rustum Roy
July 3, 1981

# Table of Contents

Preface to the Series      v

Chapter
1   PREFATORY REMARKS      1
    Scope of this Volume      1
    Disciplines Involved in the Subject Matter      2
    Special Nature of Literature Sources      3
    Authorship of This Volume      4

2   INTRODUCTION      5
    Role of the Waste-Package in RWM System: A Major Increase
      in Importance      6
    Historical Overview and Summary      7

3   WASTE PACKAGE TERMINOLOGY AND WASTE INVENTORY      13
    Inventory and Rational Categorization of Existing Wastes      14
    Solid Phase Immobilization Subsystem and the Waste Package      14
    Processing      21
    Canisters      21
    Waste Loading, Characterization and Evaluation      23

4   STATUS OF WASTE FORMS DEVELOPMENT: GLASS      26
    Glass as the (Nominal) Radiophase      27
    Glass as a Fully Engineered Waste Form (see also App. I)      35

5   ALTERNATIVE WASTE FORMS: CERAMIC RADIOPHASES      36
    Definitions, The Mineral-Model Concept      37
    Ceramic Radiophases      37
    Composition of Tailored Ceramics      39
    The Dilute-Ceramic Crystalline Solution Strategy      42
    Relative Adaptability to Compositional Change of Ceramics and Glass      44
    Processing: I. Mixing (Powder Preparation)      46
    Processing: II. Consolidation and Densification      47
    Coated Ceramic Radiophases      49
    Hydroxylated Radiophases      49
    Spent Fuel      49
    Summary on Status of Ceramic Radiophase Research      50

6   ENCAPSULATED WASTE FORMS      51
    Cermaic-Matrix Materials: The Concept      52
    Metal-Matrix Materials      55
    Glass-Matrix Materials      57

| | | |
|---|---|---|
| | Cement-Matrix Materials | 59 |
| | References | 65 |
| 7 | STABLE MINERALS, MODELS FOR CERAMICS, APPLICATION TO PARTITIONED WASTES | 72 |
| | Introduction to Concepts | 73 |
| | Scientific Foundations: mineral model concept | 73 |
| | Partitioned Wastes | 75 |
| | Conceptual Processes and Products | 76 |
| | References | 85 |
| 8 | RADIATION AND TRANSMUTATION EFFECTS | 87 |
| | Introduction | 88 |
| | Interaction of Massive Particles with Solids | 92 |
| | Effects of Radiation in Solids | 95 |
| | Discussion of Work on Candidate Waste Forms | 97 |
| | Transmutation Effects | 101 |
| | Basic Principles | 101 |
| | Chemical Mitigation Studies | 103 |
| | Summarizing Remarks | 104 |
| | References | 106 |
| 9 | CANISTERS, OVERPACK AND/OR BACKFILL | 113 |
| | Function and Terminology | 114 |
| | Canisters | 117 |
| | Materials Proposed | 117 |
| | Properties of Canisters | 118 |
| | Resource, Toxicity Considerations | 119 |
| | Overpack-Backfill | 120 |
| | Compositions Studies and Their Properties | 121 |
| | Summary | 123 |
| 10 | EVALUATION OF THE WASTE PACKAGE SYSTEM | 124 |
| | Difficulties and Limitations | 125 |
| | Properties and Processing of Waste Forms: A Comparative Summary | 127 |
| | Waste Package Evaluation: Leach Tests: Wrong Target; Wrong Conditions | 131 |
| | Previous and Current Practice of the Leach Tests; and Some Results | 131 |
| | Conceptual Errors in Using the Leach Tests to "Evaluate the Waste Form" | 133 |
| | The Alternative: The Repository Simulating Test | 135 |
| | Experimental Results of R.S.T. | 137 |
| | Some Probable Scenarios For Radioactive Waste Disposal in the United States | 139 |
| | Commercial Waste Reprocessing Plant and Repository | 141 |
| | Common Strategy Which Maximizes Safety and Minimizes Cost | 142 |

Appendix
   I    Properties of Radioactive Waste Glass in Canisters by John
           E. Mendel (Reproduced from Report PNL-2764 by permission
           of the author)     147

   II   Mineral Models for Crystalline Hosts for Radionuclides
           by G.J. McCarthy, W.B. White, D.K. Smith, A.C. Lasaga,
           R.C. Ewing, A.W. Nichol, and R. Roy     184

# List of Tables

| | | |
|---|---|---|
| 1 | Experience with Waste Solidification | 8 |
| 2 | Proposed Maximum Temperatures for Wastes in Storage Prior to 1978 | 10 |
| 3 | Inventory of Radioactive Wastes in the U.S. | 15 |
| 4 | Chemical and Physical Nature of Different Types of Wastes | 16 |
| 5 | Composition of High-Level Nuclear Wastes Showing Radical Difference Between Commercial and Defense Wastes in Mole Percent | 17 |
| 6 | Consolidation Options for Ceramics | 22 |
| 7a | Different Description of Waste Loadings Illustrating Why it is Essential to Distinguish Between Waste, Fission Product, and Radionuclide Loading | 24 |
| 7b | Relative Concentrations of Major Radioactive and Stable Nuclides in Fission Products | 24 |
| 8 | The Wide Range of RWM Glass Compositions in Use | 31 |
| 9 | Process Options in Glass Making | 32 |
| 10 | Comparison of Sets of Ceramic Phases Proposed for Commercial and Defense Ceramic Waste Forms | 38 |
| 11 | Most Important Candidate Ceramic Radiophases | 41 |
| 12 | Ceramic Matrix Waste Forms | 54 |
| 13 | Metal and Glass Matrix-Dependent Forms | 58 |
| 14 | Cement-Matrix Waste Forms | 61 |
| 15 | Comparative "Best" Reported Leach Rates for Cement-Matrix Glassy Radiophase Forms | 63 |
| 16 | Synthetic Mineral Assemblages and Repositories in the Reference Scenario | 82 |
| 17 | Yields of Stable Fission Products from Thermal Fission of $^{235}$U | 88 |
| 18 | Composition of High-Level Nuclear Wastes from Reprocessing | 89 |
| 19 | Approximate Compositions of Savannah River Waste | 90 |
| 20 | Atomic Displacements Due to Different Radiation | 91 |
| 21 | Various Kinds of Atom Displacement-Producing Techniques | 96 |

22    Relatively Abundant Fission Product Elements for Which a Valence
        Change Will Occur in a Solid Waste Form and Which Have
        $t_{1/2} < 10^6$ yr                                                    101
23    Comparative Data on Waste Forms                                         128
24    Typical Results from Repository Simulating Tests                        138

# List of Figures

I       The four major, distinctly different categories of nuclear waste.      vii
II      The total SYSTEM for Radioactive Waste Management.                       x
III     The components of the technology subsystem.                              x
IV      The research areas within each of the technology subsystem
            components.                                                          xi
V       Relationships among commonly used terms: waste form, waste
            package, and solid phase immobilization (SPI) system.                2
VIa     Spray Calciner/In-Can Melter Schematic.                                 29
VIb     Spray Calciner/Ceramic Melter System Schematic.                         29
VII     Ionic Radius Plot.                                                      42
VIIIa   No compositional change within the triangle can change the phases.      45
VIIIb   Comparison of compositional range of forgiveness in two
            target compositions.                                                45
IX      Showing relation of unpartitioned and partitioned approaches
            to solidification.                                                   77
X       Various levels of partitioning.                                         78
XI      Reference process for each waste fraction.                              83
XII     Alternative overpack arrangements.                                     115
XIII    The principle of the chemical (thermodynamic) overpack.                122
XIV     Role of technology in total Radioactive Waste Management.              130
XV      Role of water in repository failure.                                   134
XVI     Comparison of leach test and repository conditions.                    135
XVII    New repository simulating test.                                        137

# Chapter 1.   PREFATORY REMARKS

## Scope of This Volume

This monograph deals with one component of the technological subsystem for radioactive waste disposal, the waste package--or more comprehensively, the solid phase immobilization system.  The preceding preface to the series has located this waste package component in the technological system in the following simplified scheme:

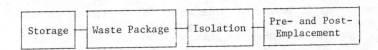

In this monograph we will be concerned with most of the aspects of research being done in the fields listed below:

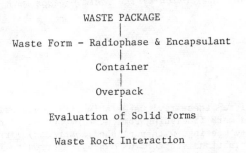

We restrict our concern to what has been variously referred to as the (Radionuclide) Immobilization System, or what we shall henceforth call The Solid Phase-Immobilization (SPI) Subsystem, which *includes* The Waste Package, "Containers" and "Overpack."  The diagram clarifies the terminology we will use:

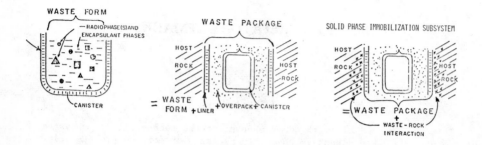

Fig. V. *The sketches show the relationships among commonly used terms: waste form, waste package, and solid phase immobilization (SPI) system. The term* radiophase *refers to the solid entity or phase in the Gibbsian thermodynamic sense which contains any radionuclide. Each crystalline ceramic (artificial mineral) or glass with its individual structure constitutes a separate radiophase. The term* encapsulant phase *refers to all inert or non-radionuclide containing phases, which surround or physically isolate radiophases. The* waste form *is any combination of radio and encapsulant phases. Waste* package *includes the waste form in a canister and surrounded by overpack or backfill SPI system includes the additional layer of the chemical interaction of the waste package with the host rock.*

We have restricted these reviews of scientific data principally to *published*, refereed work (including some work presented at national professional society meetings with published abstracts). Problems related to technological or national objectives *may* by referred to from non-traditional sources: agency reports, Congressional hearings, etc. However, we do *not* aim to be exhaustive or encyclopedic but, rather, negentropic, organizational, evaluative. The goal is to provide the nonspecialist in materials science with a reliable, technical, but not recondite, state-of-the-art survey of *waste solidification* science and technology.

## Disciplines Involved in the Subject Matter

A glance at the listing of topics listed in the previous paragraph would immediately indicate that the subject matter content of what will be

treated herein is not nuclear physics or nuclear engineering. The
discovery of nuclear reactions and their utilization to produce weapons and
civilian power was appropriately carried out by physicists and nuclear
engineers. However, this "back end of the fuel cycle," the garbage
disposal system, calls for very different expertise. Indeed this major
change in the discipline base needed to solve waste disposal problems, was
in no small measure responsible for its neglect. Physicists, nuclear
engineers and chemical engineers are not familiar with the science of
designing and actually making solids which will retain the radionuclides in
maximally insoluble phases. These disciplines do not treat the reaction of
solids within this earth for times ranging from 1000 years to millions of
years.

Making solid materials is the province of materials science and
engineering. The conditions for reaction of such materials in the near-
surface part of the earth is the province of the experimental geochemist.
The reaction of the inserted waste package with its host rock and any
intruding fluids is part of mineralogy and petrology. In a remarkable
instance of technical shortsightedness (except for the quite recent past),
virtually no members of any of these large scientific and engineering
communities was involved in the development of the technology of radioactive
waste disposal. That situation is now changed and this volume will, of
course, treat the subject matter from the viewpoint of the present status
of materials science and the experimental geological sciences.

## Special Nature of Literature Sources

The literature in a field such as radioactive waste disposal is most
unusual in three respects. First, it exists largely *only* in reports (many
only internal) of various national atomic energy agencies or their
equivalent, and much of it is slanted towards the solution of engineering
problems. Second, most of the explicitly designated normal peer-received
journal literature in RWD is hardly three years old. Third, the relation
of the solidification work to the technological and, indeed, the larger
RWM system is treated by a very wide spectrum of authors. Many of these
are not technically trained; but even among scientists writing in the
field, the normal canons of scientific journalism are often violated. Thus
authors totally unfamiliar with materials sciences or geological sciences
publish papers in the field on the materials or geo-aspects of the subject
in specialized RWD literature where inadequate reviewing leads to a great
deal of confusion. We make the reader aware of this unusual nature of the
literature since it is easy to be misled by a great diversity of uninformed
opinions of experts writing outside their specialties. In the sections
below we try to provide a guide to the reliable literature in this field.

### PSCRWM as Baseline on Overview Literature

In 1976, the National Academy of Sciences under contract to the Nuclear
Regulatory Commission started a study on the solidification of radioactive
waste, under its Standing Committee on Radioactive Waste Management. The
report (1) of this committee is perhaps the first *policy* document to consider

the question of the many alternatives within the SPI subsystem as we define it here. It contains the only serious consensual evaluation of the efficacy of various waste forms for different wastes. Following the issuance of the document, although not wholly as its consequence, U.S. national policy itself and later other material groups moved to provide considerable financial resources for research on some of these solidification alternatives.

## Professional Society Meetings

Over two decades, some dozen major technical meetings of professionals, under auspices such as the IAEA, the American Nuclear Society, A.I.Ch.E., etc., had been held on the subject of waste management. Such meetings typically focussed either on the management, biohazards, chemical engineering, or the geological isolation aspects of the total system. The new surge in interest in the *materials science* relevant to the problem was recognized by the Materials Research Society, which in 1978 started an annual series of international symposia on the materials science and technology of RWM in conjunction with their annual meetings. In 1979 the American Ceramic Society also ran a symposium on the ceramic and glass aspects of the problem. The increase in research support will result now in a steady stream of papers in the regular journal literature and the task of judging the value of such fragments of knowledge in a highly integrated field will not be easy for the reader. The published proceedings of the continuing series of Materials Research Society conferences provide a new, more or less complete, source of significant data for keeping abreast of waste package science and technology. In our own judgment, however, since the funding and actual research activity in most laboratories is so new, much of the work so far reported is in the nature of progress reports. Exceptions exist, of course, in laboratories working in the same area for a long time, for instance in the work on glass from the Battelle Pacific Northwest Laboratories, the concrete and grouting process at Oak Ridge, and on ceramics at Penn State and the Sandia Laboratories. These results, whenever appropriate, are incorporated into the following sections. The literature, especially the report literature, is expanding very fast, and hence the survey must be cut off at some date. This date was approximately the end of 1980.

## Authorship of This Volume

From the Table of Contents it will be clear that while the author has been responsible for the architecture of this book and writing of most of it, two chapters and one appendix have been contributed by his Penn State Colleagues as indicated. In addition, we are grateful to Dr. J.E. Mendel, one of the pioneers of glass waste research in the U.S. for permission to reprint one of his reports on the properties of glass waste forms as Appendix II.

# Chapter 2.   INTRODUCTION

## *Synopsis*

The role of the solids which incorporate, immobilize and retain the radionuclides has in the last five years undergone a radical increase in importance.  For 25 years emphasis was (100%) on safety via isolation of the transporting liquids from the biosphere by judicious selection of the geology.  The perception today is that there is a two-pronged series approach to preventing radionuclides from reaching the biosphere:  (a) by maximizing the insolubility or unreactivity of the solids which contain them, and (b) by minimizing the rates of liquid transport to the biosphere.

Contrary to popular opinion much greater progress has been made on waste solidification (and disposal) than believed.  Thus several hundreds of megacuries of solid waste form have been made and finally disposed of in the U.S. (at Oak Ridge) and in the USSR.  A second great surprise to most technical readers is the fact that vastly more radioactive waste has been solidified into ceramic waste forms at one location in the U.S. (Hanford) than the worldwide amounts made into glass.

Part of the reason for the absolute necessity for having an array of options of solid waste forms is that even the existing so-called high-level wastes are so different in composition and activity (at least two orders of magnitude).  Moreover, as the design temperatures and locations of the repository change, the suitability of a particular waste form also changes. It is shown that the changes in design temperatures and wetness conditions have demanded changes in the solid waste forms for the new conditions.

# Role of the Waste-Package in RWM System: A Major Increase in Importance

With the wisdom provided by hindsight one can see that the principal reason for many of the shortcomings in research strategy in the RWM field was the absence of a fundamental conceptual understanding of the *total* radioactive waste disposal *system*. And as a result, the role of the solid phases responsible for immobilizing the radionuclides within the system was not clearly grasped. In the classical era of radioactive waste management (approximately, during the lifetime of the Atomic Energy Commission), the development of the "system concept" was inhibited for two reasons. Firstly, there was a lack of urgency in designing the ultimate disposal, and secondly, the focus was too strongly skewed towards the geological isolation; so much so, that the solid phase was considered merely as a vehicle for transporting the radionuclides to the isolation site. This picture has changed substantially in the last three or four years. A rather generally accepted description of the *total system* has been provided by the National Academy of Sciences Study on Waste Solidification (1), and later developed more explicitly by Roy (2) in the overview, introductory paper at the first International Symposium on Radioactive Waste Management at the Materials Research Society. The scheme on page -- shows the components of the national system taken from these papers, and also lists just the headings of the major subsystems for easier identification of the areas of research with which we will be concerned.

Figure III (see general preface) provides a simple diagrammatic way of assessing the *potential* role of the waste package (or solid phase immobilization) in the total system. In the past, and as recently as 1978, many systems were designed or evaluated assuming that the waste package barrier effectiveness was zero (i.e., that the solids dissolved totally). This threw the entire burden on the isolation subsystem. While considerable research has been done on the isolation part, it has two major drawbacks. It is most susceptible to socio-political objections, and it is beyond verifiability. Recently, however, the waste package research (such as that reviewed herein) of the last few years, has raised the possibility of the obverse of the above assumption that the isolation effectiveness can be ignored. The socio-political significance of the achievement of a near "zero-release" radionuclide immobilization waste package has been emphasized elsewhere (3). If the goal of the SPI research is seen as the design and production of a "package" which by laboratory tests can be shown to release essentially zero radionuclides under the conditions of temperature, pressure, Eh, pH, etc., which any plausible geological or seabed repository may determine, then the options available to the national policy maker can be dramatically increased. Indeed, it now appears likely on the basis of recent results that the solid phase immobilization subsystem can provide both a sufficiently durable, laboratory-verifiable component of the total system to *fulfill* such a goal.

*Thus, the "solidification" and "isolation" links in the RWM systems stand today as equal partners in providing the assurance of the public's protection against possible radionuclides' release to the biosphere. The solid-phase immobilization system has the advantage of being testable in the laboratory.*

## Historical Overview and Summary

Hot Solidification into Many Forms Already Achieved

*Contrary to a large number of statements, both waste solidification and even solid waste disposal have been carried out (without a single fatal radiation-connected mishap) at full scale engineering levels within the U.S. for over a decade.* Oft-repeated, totally unsubstantiated, claims and allegations to the contrary are made principally by individuals and organizations opposing the development or deployment of nuclear power.

The claims are at different levels of generality:

(a) The problem has never and can never be solved (mainly in reference to "millions of years").
(b) No actual disposal ever done.
(c) No hot engineered system full scale repository working.
(d) No hot engineered solidification ever done.
(e) No hot solidification done.
(f) No solidification done.

It will be seen that except for (c), all these claims are quite incorrect. However, if only because such statements are not vigorously rebutted, it often appears that even strong advocates of nuclear power are unaware of the status of the technology and engineering of *"waste-solidification"* (*not* ultimate disposal).

Table 1 shows a list of the status of *hot, engineered,* solidification systems which are or have been fully functional in the U.S., Europe, U.K., and USSR. The amounts of waste are order-of-magnitude estimates but are sufficient for our review purposes. We emphasize for the reader that *both the natural mineral-matrix composites in the USSR and the cement-matrix composites at Oak Ridge represent final engineered* <u>*disposal*</u>. By 1981, Oak Ridge will be able to dispose of 10 megaCi in the new facility under construction.\* It is virtually certain\*\* that the USSR is continuing its disposal of substantial quantities of "high-level" wastes in a similar manner.

Table 1 also shows that major waste *solidification*, has been conducted hot, and at full scale engineering, at two sites in the U.S. In both cases the solid form has been a ceramic--in one case fully crystalline powder compacts encased in metal canisters, which has been successfully stored, as well as transported, and in another fine (presumably) cryptocrystalline powder stored in stainless steel bins--available for further processing.

---

\*It is estimated by Oak Ridge personnel that a single grouting well *could* dispose of 100 grout sheets, each sheet being formed of 100,000 gallons of grout @ *say* a loading of 10 Ci/gallon, for a total of about 100 megaCi. Thus, where the *geology is favorable*, concrete grouting is probably more than two orders of magnitude less expensive than any other method of *solification and disposal*.

\*\*Personal communication to R. Roy by V.I. Spitsyn, Institute of Physical Chemistry, Moscow.

TABLE 1

Experience with Waste Solidification

| | | Total radionuclide inventory of solidified waste (in megacuries) | Engineering Development Status | Back-up research data on process, properties, etc. |
|---|---|---|---|---|
| HOT SOLIDIFICATION ONLY | Crystalline (= ceramic) Radiophases | 400 | Successful fully engineered system (Hanford) | small |
| | Cryptocrystalline Radiophases (calcines) | 100 | Successful continuous operation for 8+ years (ICPP) | small |
| | Glass (+ crystalline) radiophases) | ∿30* | Hot test runs in U.K. + U.S. Hot *plant* in France | very large |
| | Metal matrix composites | ∿1 | Hot pilot plant in Belgium | moderate |
| FINAL DISPOSAL COMPLETE | Cement-based composites | ∿1 | Fully operational in U.S. (Oak Ridge) | small |
| | Natural mineral matrix composites formed in situ pumped into porous sandstone | ∿100 ++ | Hot full scale disposal in USSR | large |

*Including a total of 24 MCi in France estimated by G. Sombret: letter to R. Roy (1980).

*The status of the SPI subsystem has been changed due to the realization that many solid waste forms have been developed and successfully engineered. In terms of engineering experience to date worldwide, the rank order of these forms by amount consolidated is: (unconsolidated) crystalline ceramics; (unconsolidated) cryptocrystalline ceramics; mainly glass (+ ceramics); cement-based composites; natural mineral composites.*

---

Table 1 also shows, surprisingly, that as yet there is relatively little hot continuous experience with the glass waste form. However, this is partly compensated for by extensive R&D on this form all over the world, and the commitment to full scale production in France.

Our first conclusions therefore are:

1. A. *Hot, waste solidification has been successfully engineered in the following solids, in descending order of experience*
   a. Mineral-matrix . . . . . . . . . . . (Low-temperature forms)
   b. Crystalline ceramic forms }
   c. Cryptocrystalline ceramic forms } . . . . Moderate temperature forms
   d. Glass and glass-in-metal matrix forms} . . High temperature forms
2. B. *One waste solidification-and-disposal process has been successfully demonstrated in two locations.*
   a. On-Site Pumping or Grouting of a Cement or Mineral Matrix Into Carefully Selected Geologic Horizons

## Revision of Design Objectives

There has recently been a second major change in the technological strategy for the total system. For many years most nations concentrated on developing a system, the components of which had been chosen without appropriate conceptualization of the possible *failures* in the system. Thus the reference system focussed as late as some two or three years ago on making a (zinc) silicate glass, sealing in a stainless steel canister, and burying the latter in salt. The systems were designed to operate--when successful-- under *dry* conditions. The limits on the age of the waste were not carefully specified, nor was the concentration of the radionuclides. Although various small research programs had studied alternatives, it is fair to state that *up to 1977*, the national U.S. program had opted to *develop* a single waste form--a "glass" for "ultimate" disposal. The official design temperatures which may be found in the literature--in *the light of hindsight*--often sound rather high because we forget that they were designed to operate dry. Temperatures run typically in the 300°C-400°C range (4) for both the geological storage and surface storage options (see Table 2). But in one failure mode (accidental burial of a retrievable surface storage facility [RSSF] canister to 1.5m depths in surface layers), the temperature was calculated to reach 930°C (5). Even in the regular performance of the proposed RSSF with no air cooling, the *surface* canister temperature was calculated to be ∿800°C (see Table 2).

In 1978 a paper by McCarthy et al. (6) first showed experimentally some of the consequences of the possible *failure* of the isolation system to keep the reference form (glass) *dry* at the then current design temperatures. Under the simulated mild hydrothermal conditions, typical waste glass was nearly totally transformed in a matter of weeks. An indirect result of this work was to intensify the research on waste solidification in two directions. First, was the change of design objectives to much lower center line temperatures of the solid, to be achieved by either dilution (or aging). Second, ·was tne search for alternative solids which could withstand higher temperatures and "wet" conditions.

## The "Reference" Waste Form - Glass

Readers of this report will be familiar with the fact that to a very large extent, "glass"-making was synonymous with waste solidification in all the nuclear establishments of the world except the USSR for two decades. It is not clear exactly how or where the leaning to glass originated. The regular journal literature and the normal professional societies' meetings carry hardly any papers explaining this choice. Nevertheless, numerous specialized meetings devoted to radioactive waste management reported on the development of glass as a waste form in Harwell, in the U.S. at the Battelle Pacific Northwest Laboratories, and in France. Two fine summaries of the status of research on the glass waste form have been provided by J.E. Mendel and co-workers (7a,b). Several large (∿30cm x 2.5m) steel canisters of glass had been made, some full scale. A variety of quite different glass compositions had been developed incorporating 10-30% fission products. All these glasses were relatively resistant to "leaching" by water at 25-100°C ($10^{-5}$ gm/cm$^2$ per day), although the testing procedures used worldwide were rather crude, by modern scientific standards. The glass-melting process developed in different countries also differed

TABLE 2

Proposed Maximum Temperatures for Wastes in Storage Prior to 1978

*Sealed Storage-Cask Component Temperature Data*

| | | With Air Flow Temperature, °F | | | No Air Flow[3] Temperature, °F | | |
|---|---|---|---|---|---|---|---|
| | | Operating | Limiting | Factor[4] | Operating | Limiting | Factor[4] |
| 5 kw[1] | Canister | 470 | 800 | 1.7 | 810 | 1500 | 1.8 |
| | Cask | 245 | 550 | 2.2 | 700 | 1200 | 1.7 |
| | Concrete, inside surface | 155 | 500 | 3.2 | 650 | 900 | 1.4 |
| | Concrete, outside surface | 84 | 500 | 5.9 | 120 | 900 | 7.5 |
| 10 kW[2] | Canister | 650 | 800 | 1.2 | 1300 | 1500 | 1.2 |
| | Cask | 390 | 550 | 1.4 | 1200 | 1200 | 1.0 |
| | Concrete, inside surface | 255 | 500 | 2.0 | 1190 | 900 | 0.8 |
| | Concrete, outside surface | 95 | 500 | 5.2 | 135 | 900 | 6.7 |

NOTE: All data is for PWR-U-type waste, 2-inch-thick cask, 6-inch annulus.
[1]For 5 kW, Air flow = 3 fps, 600 CFM, Ambient Air = 80°F, Exit Air = 115°F.
[2]For 10 kW, Air flow = 4.2 fps, 830 CFM, Ambient Air = 80°F, Exit Air = 190°F.
[3]For the No air flow data, the limiting temperatures used for the canister and the cask are based on 240 hours.
[4]Factor = Limiting temperature/operating temperature.

*Surface Storage Concept: Data from Interim Storage of Solidified High Level Waste, National Academy of Sciences, Washington, DC (1975). See also Tables 1, F-2, 3, and page 26 in same report.*

Maximum Temperatures* and Gamma Doses in Rock Adjacent to HLW

| Rock | Maximum Temperature (°C) | Maximum Gamma Doses (10[10] rad) | | |
|---|---|---|---|---|
| Shale | ∿300 | 2.8 at 100 yr | 3.1 at 1000 yr | 4.3 at 10[7] yr |
| Shale and other rock | ∿400 | " | " | " |

*\*Geological Storage Concept: Data from Glenn Jenks, National Waste Terminal Storage Program Conference on Waste-Rock Interactions, Report Y/OWI/SUB-77/14268 (July 6-7, 1977), The Pennsylvania State University, University Park, PA 16802.*

significantly, from in-can melting at Harwell, a small continuous-flow electrically-heated tank at Marcoule, to bead-making in Mol, Belgium. The question of the possibility of improving on a glass as the choice both with respect to waste form *and* the process used, was seriously raised only as new groups of scientists, especially those concerned with materials and geosciences, became involved in the issue.

The Resistate Mineral-Model:  Conceptual and Experimental Advances

It is self-evident to anyone trained in the geosciences, familiar with postwar experimental petrology, that the existence of multiphase mineral assemblages which are either (thermodynamically) stable or persistent for geological times *at elevated water pressures and temperatures* constitutes excellent proof of the immobilization in mineral crystals of dozens of radionuclides for "geologic times."  Those assemblages were probably stable when formed in most cases, but what is significant are the evidences and arguments long since accepted by sedimentary and metamorphic petrologists of the metastable persistence--and hence lack of weathering and dissolution--of a certain set of mineral phases under many geologic environments encountered in the upper kilometer or two of the earth's surface.  Furthermore, the incorporation of several percent of actinides and certain daughter radionuclides in some of these natural phases with no apparent change in survivability provides evidence that the various radiation processes make no major difference in the properties relevant to nuclear waste isolation.  One of the most significant aspects of the mineral model is the "experimental" demonstration of the survival of such phases in the field over millions--indeed in some cases, billions of years.

Such facts about crystalline ceramic materials in general led L.P. Hatch of Brookhaven to write in detail of the advantages of using such mineral assemblages for radioactive waste solidification rather early in the history of solidification (1953) (8).  Very early encapsulation work with an inert matrix of powdered mica and glass frit, was done by Evans and Macdonald*, but it does not seem to have been followed up or published in the open literature.  In 1963, Arrance was issued a U.S. patent for a "ceramic sponge" process (9).  In 1972, Isaacson and Brownell (10) revived the same arguments for studying mineral-based polycrystalline ceramic waste forms.  However, systematic experimental research explicitly based on the fundamental petrologic stability of a specific set of crystalline mineral radiophases, and their experimental testing under those conditions was only started under an NSF grant** by McCarthy in 1971.  This work on mineral-modeled ceramic phases of all kinds and use of state-of-the-art ceramic processing became the nucleus of the Penn State work funded under the Battelle Pacific Northwest Laboratory's program on alternative *ceramic* (see discussion of terms below) waste forms, and reported on in a series of papers by McCarthy, starting in 1973 (11a-e).  The conceptual model was detailed by Roy in 1975 (12).  Additional support for the resistate mineral concept was provided by Ringwood in 1978 (13a,b), who added the proposition of dilution of the radionuclides in the host structures to increase leach resistance still more, since cost in his view was not important.  Details on the specific phases and processes are presented later.

---

*E.J. Evans and J.F. Macdonald (1959), Hot-Pressed Silicate Materials for the Fixation of Fission Products, Rept. CRER, A.E. Canada Ltd., Chalk River, Ont., p. 795.

**The abstract of the research objectives stated in part:  "The important criterion of nonleachability will be tested for selected light rare earth silicates, aluminosilicates, aluminates, phosphates, niobates, and tantalates at simulated deep burial site conditions."

## Other Waste Forms Emerge

Meanwhile, developments of alternative SPI subsystems were occurring
elsewhere, although without as much direct attention to the primary radiophases. Research in the Soviet Union (14) had focussed on the use of deep
well injection of "low-level" liquids or slurries into porous sandstone
media. At Oak Ridge National Laboratory (15), a similar process consisting
of injecting a cement + waste slurry + additives, grout into hydrofractured
shales at depths of ∿500m had been developed over the years. Since 1966
the facility has successfully *ultimately* disposed of nearly a megacurie of
"intermediate" level wastes. In both these cases, the resultant "solid
waste form" consisted of largely unknown radiophases, microencapsulated by
the phases of hydrated cement in the Oak Ridge case, or the quartz and mica
in the original sandstone in the USSR case. Although not generally recognized, the cement-waste composites are the most developed examples of a
rather different *class* of waste forms which rely on microencapsulation
rather than insolubility of phases alone. Extensive recent laboratory
studies on encapsulation in essentially pure $TiO_2$ have been reported by
Schwoebel and Johnstone at Sandia (16a,b), and by Westermark (17).

In 1977, the NAS Radioactive Waste Management Committee, aware of the
change in the importance of the waste form, for the first time in two
decades established a Panel on Waste Solidification. The Panel's report
(1) clearly validated the concept of the viability of *many alternative
solid forms*, and pointed to the advantages of *diversity in waste-forms for
different cases*, such as old defense and fresh commercial wastes.

All these developments formed the background to the enormous increase
in attention and funding to alternative waste form research at the beginning
of FY'80 in the U.S.

However, before we can discuss the science and technology involved in
the solid phase immobilization subsystem, it is essential that we attempt
first to establish a common terminological base. Great confusion has been
caused by the use of different names for the same entity, by misuse of
universally accepted scientific terms, and by devising of *"brand"* names to
describe specific compositions of waste forms.

# Chapter 3.   WASTE PACKAGE TERMINOLOGY AND WASTE INVENTORY

## *Synopsis*

*We present here definitions of terms used in this and related reports. Without such a glossary the serious reader is at sea since different groups attach different meanings to the same term.*

*The subsystem for the immobilization of radionuclides in solids is divided into its components such as radiophase waste form, waste package, etc. and the relations between such terms clearly described. Terms such as "waste-loading," and the various means of "evaluation of waste-forms" are extremely important, and the definitions of these are treated in some detail.*

# WASTE PACKAGE TERMINOLOGY AND WASTE INVENTORY

We have already referred to the fact that the early history of the field involved professionals from disciplines outside those most relevant to the subject matter. As a result, there has also been a regrettable tendency for those not trained in the disciplines to use terminology incorrectly, and invent new terms even where these are quite unnecessary. Whenever relevant we will use the standard terminology of materials science, processing, and characterization. Thus where applicable we have followed the National Materials Advisory Board's major studies on Materials Characterization (#229-M) and Ceramic Processing (#1576) (18a,b). We turn first therefore to the definition of terms.

## Inventory and Rational Categorization of Wastes

In the preface to the series we alluded to the four very different categories of waste which exists. Not only are these radically deficient in nature but also in the amounts which exist. Table 3 brings together the information regarding where the amounts are located and the amounts which exists of three of the four major waste categories (see Fig. I). Mill tailings, not listed in the table, are scattered near mining or milling sites in the U.S. mountain states and other parts of the world, in amounts measured in the tens of millions of tons.

It has become evident that one of the major difficulties in proper discussions of the RWM issues, is the tendency to treat all "high-level" wastes as a single "commodity." The compositional differences between different classes of *waste* are so large that such gross treatment causes enormous distortions.

In Table 4 we also summarize the general nature of these wastes-- whether a clear acidic liquid stream or refractory oxides in metal tubes to emphasize these differences. And in Table 5 we carry the description to a finer level by showing and comparing the chemical composition of typical commercial waste with typical defense waste.

The importance of these distinctions is substantial in waste package considerations because of a basic crystal chemical principle. In the high concentration wastes the structure and hence the properties of resultant waste forms are based on the phases formed and hence "controlled" by the radio-ions population will be accommodated in crystalline solution in host structures which are predetermined by nonradioactive ions. The same holds for dilute noncrystalline, or cryptocrystalline, host structures. Further- more, in these wastes, the inert encapsulant phase properties, if such a phase is present, will play a major if not dominant role. Thus the solidification strategy is *radically different* depending on whether one chooses the low concentration or high concentration approach.

## Solid Phase Immobilization Subsystem and the Waste Package

The isolation of the wastes in a undersea or land-based mined repository achieves the separation of radionuclides from the biosphere by

Table 3

Inventory of Radioactive Wastes in the U.S.[a]

| Present Location | Present Form | Quantities | | Storage Mode |
|---|---|---|---|---|
| | | $10^6$ gallons | $10^6$ Curies | |
| ICPP[b] | Calcine (solid) | 0.45 | ⎫ | in 4 sets of SS bins |
| ICPP | Acidic liquid | 2.5 | ⎬ 85 | in 11 SS tanks |
| ICPP | Acidic waste generation rate | 0.5/yr | ⎭ – | ---- |
| Hanford[c] | Liquid (alkaline) | 8 | ⎫ | in 168 tanks |
| Hanford | Salt cake | 30 | ⎬ 190 | |
| Hanford | Sludge | 12 | ⎭ | |
| Hanford | Capsules (solid SrF$_2$, CsCl) | 240 capsules* | 350 | water cooled pools |
| SRP[d] | Liquid (alkaline) | 14 | ⎫ | in 33 tanks |
| SRP | Salt cake | 7 | ⎬ 570 | |
| SRP | Sludge | 2 | ⎭ | |
| SRP | Liquid waste generation rate | 1.5-2.0/yr | ? | ? |
| NFS[e] | Liquid (acidic) | 0.01 | 2.1 | |
| NFS | Salt/sludge/ liquid (alkaline) | 0.6 | 64 | |
| Various[f] sites | Spent fuel | 6,500 MTHW[h] | 2,800 | water cooled racks |
| Various[g] sites | $^{85}$Kr (gaseous) | -- | 3.0 | released |
| | $^3$H (gaseous) | -- | 0.5 | released |
| | $^{14}$C (gaseous) | -- | $7 \times 10^{-4}$ | released |
| | $^{129}$I (gaseous) | -- | $1 \times 10^{-5}$ | released |

[a]This inventory includes high level wastes and gaseous radionuclides that are likely candidates for containment in solid matrices. Data obtained through surveys conducted for DOE by Dr. K.K.S. Pillay (Penn State University).
[b]Idaho Chemical Processing Plant, Idaho National Engineering Laboratory, Idaho.
[c]Hanford Reservation, Richland, Washington.
[d]Savannah River Plant, Aiken, South Carolina.
[e]Nuclear Fuel Services, Inc., West Valley, New York.
[f]Spent fuel storage pools derived from about 70 light water power reactors now in operation and in other centralized storage areas.
[g]All DOE Research, Development and Production sites plus all U.S. nuclear power plants.
[h]Metric tonnes of heavy metal as of 1979, with an annual addition of ∿2000 MTHW from the commercial LWR's now in operation.
*Capsule size ∿1m x 10cm.

the physical prevention of the *transport of liquids* from the proposed repository. However, before any ion can be made available for such transport it must be dissolved (leached) out of the solids into which the radionuclides were engineered. This step of molecular engineering, or materials design, synthesis and processing leading to a certain series of solids which retain the radionuclides within the waste package, is the subject of this entire volume, and we start with definitions regarding it.

Table 4

Chemical and Physical Nature of Different Types of Wastes

| Total Waste Composition | Nature | Radionuclide Concentration |
|---|---|---|
| Nuclear Weapons Program =(LOW-CONCENTRATION WASTES) | Solid + liquid *slurries* mixed with inert salts and various inert precipitates, in very large steel tanks | Concentration of radionuclides in original solids (<1%) |
| Wastes from reprocesses (Commercial reactor fuel)    =(HIGH-CONCENTRATION WASTES) | a. *Liquid* effluent stream from re-processing plant (<10 years out of reactor) (none exist in U.S.)   b. Solid + liquid phases (as sludge). Only in TMI, Harrisburg | Concentration of *radionuclides* in calcined solids: ∿10 wt % (excl. An*) |
| Mixed case | Similar to above at N.F.S., West Valley, NY | |
| SPENT FUEL | Refractory-Oxides (+ gas) encased in metal rods | 90% radionuclides (including $U^{238}$) 80-100% radiophases (including $UO_2$ containing Ln**, Sr, etc.) |

*An = actinide     **Ln = lanthanide or rare earth

TABLE 5

Composition of High-Level Nuclear Wastes Showing Radical Difference
Between Commercial and Defense Wastes
in Mole Percent

| Component | Commercial | | Composite Defense + Savannah River Sludge |
|---|---|---|---|
| | PW-4b | PW-7 | |
| Rare earths | 26.4 | 30.9 | $\sim$1.25 |
| $ZrO_2$ | 13.2 | 9.2 | $\sim$0.3 |
| $MoO_2$ | 12.2 | 8.2 | $\sim$0.005 |
| $RuO_2$ | 7.6 | 5.1 | $\sim$0.2 |
| $Cs_2O$ | 7.0 | 4.7 | $\sim$0.03 |
| $Fe_2O_3$ | 6.4 | 8.7 | 36.13 |
| Pd | 4.1 | 2.8 | $\sim$0.01 |
| SrO | 3.5 | 2.3 | $\sim$0.07 |
| BaO | 3.5 | 2.3 | $\sim$0.2 |
| $[PO_4]$ | 3.2 | 8.7 | $\sim$0.2 |
| $U_3O_8$ | 1.4 | 9.6 | $\sim$3.26 |
| $Rb_2O$ | 1.3 | 0.9 | ? |
| $Na_2O$ | --- | 0.9 | $\sim$6.01 |
| $Am+Cm+PuO_2$ | 0.2 | 0.2 | (see $U_3O_8$) |
| Others, e.g. | 9.8 | 5.5 | |
| $Al_2O_3$ | | | 28.2 |
| $MnO_2$ | | | $\sim$9.94 |
| CaO | | | $\sim$2.69 |
| NiO | | | $\sim$4.47 |
| $SiO_2$ | | | $\sim$0.85 |
| Zeolite | | | $\sim$8.93 |
| Hg | | | $\sim$1.5 |

PW is U.S. Energy Research and Development Administration
shorthand for Purex process Wastes.

PW-7 contains an excess of gadolinium used as a neutron
absorber plus more iron, $[PO_4]$, uranium, and some sodium.

The minor elements designated with the $\sim$ prefix are given
as the average of the range.

In Fig. V, we presented a schematic drawing of the total subsystem and this should be consulted in connection with the discussion. We explicitly define the SPI subsystem as including the following four 'barriers,' starting from the radionuclides and working outwards physically, and the waste rock interaction occurring next to the waste package.

$$
\text{Solid Phase Immobilization Subsystem} =
\begin{cases}
\text{Waste Package} = \begin{cases}
\text{Waste Form} = \begin{cases}
\text{Radiophases (contain radionuclides)} \\
+ \text{ possible} \\
\text{Encapsulant phases (contain no radio-nuclides)}
\end{cases} \\
+ \\
\text{Canister} \\
+ \\
\text{Overpack (=backfill)}
\end{cases} \\
+ \\
\text{Waste Rock Interaction}
\end{cases}
$$

The *waste package* is the entire entity which is inserted into the host rock. The *waste form* is an often-used term, and as the illustration shows includes the *radio-phase* and *encapsulant phase*(s) taken together, or that which is inside the container (if there is one).

We note that in any particular real example any one or two of the latter barriers may not be used. *Indeed, it must be remembered that only the primary radiophases (even if we do not know what they are) are part of every waste package.* The following examples of several different total waste packages will illustrate the use of these terms.

In order to understand how each waste package functions it is essential to understand the crystal chemistry and physics of each of these materials and the roles they are designed to, or can, play. For illustration let us take several examples of total systems, and note the materials used in each stage.

*Example (1).* Borosilicate glass beads in a lead metal matrix contained in a steel canister and emplaced in a salt repository with a crushed salt backfill. The radiophases are the noncrystalline glass or glasses (when phase separated) plus crystalline metal and oxide inclusions. Lead is the encapsulant phase and steel the container. There is no overpack.

*Example (2).* Tailored ceramic monolithic cylinders (10cm x 10cm) prepared by hot pressing, stacked in a steel canister, make up the radio-phases. The ceramics could have phase assemblages tailored for high radio-nuclide loadings, such as those developed by McCarthy et al. (11). Typical radiophases would be [pollucite], [fluorite], [spinel], [perovskite], [monazite], etc. Alternatively, the radiophases may be those that accept only low loadings, such as those of Ringwood (13a). The many radionuclides will be partitioned into one or more of the preselected phases. In these ceramics, the nanostructure of the radiophases is known or can be determined in detail by available tools. There is no encapsulant phase in this example.

The hot-pressed ceramics could be placed, let us say, in a basalt repository encased in a metal container, surrounded by a tailored overpack of, say, clinoptilolite + mordenite. These two zeolite minerals would have very high adsorption for most of the radioions which escaped the container.

*Example (3).* A third example fully worked out, on paper, is the Swedish KBS-II plan. This proposes to use as the only radiophase a (French) borosilicate glass with some 9% of fission product oxides which have been about 50 years out of the reactor. The radiophase would have no surrounding encapsulant phases, but would be placed in a heavy walled copper canister. This would be surrounded by a tailored overpack of bentonite + quartz [and $Fe_3(PO_4)_2$ to control Eh], designed chiefly to keep water away from the canister, in the granite repository.

*Example (4).* A fourth example which has been developed fully through the laboratory scale by the Swedish ASEA company is as follows. The radiophases are the fission product-rich, tailored ceramics of McCarthy; the encapsulant phases are mullite (+ alumina?). The waste form is hot isostatically pressed. The container is a 3m x 0.5m cylinder, hot isostatically formed and sealed, made of "sapphire" with a 10cm wall. This container could be surrounded by a zeolite overpack in a granite repository.

*Example (5).* The last example is the one which is the most developed and least discussed in the literature. This is the Soviet and Oak Ridge final disposal systems actually in use. Here, as yet, no published work can tell us what the radiophases actually are: One can estimate that the clay phases in the additives as well as many of the hydrated phases of cement* become dilute radiophases*. But many of the radionuclides will form very small amounts of other phases which have not yet been identified. Surrounding the radiophases will be a mass of encapsulant phases consisting of the unreacted anhydrous cement phases*, etc., and various hydrates which do not incorporate significant amounts of radioions. In the USSR scheme, the radiophases are almost certainly expanding layer silicates plus the unknown minor phases while the encapsulant phases are the quartz, mica, and other minor constituents of the sandstone. In both these solidification and *disposal* schemes, neither a container nor an overpack is utilized.

<u>Classes of Radiophase(s)</u>. Three classes of materials are *relevant in RWM:*

*(a) Ceramics (inorganic, polycrystalline, polyphasic* (usually insulator) materials): NO OTHER TERM such as artificial minerals, synroc, supercalcine, has ever been necessary; 'ceramics'** should be used, as the single generic term for all crystalline** radio and/or encapsulant phases. Most

---

*These complex structures are often designated by their $CaO;Al_2O_3:SiO_2:H_2O$ ratios using C, S, H, as symbols for the oxides. Thus, the chief hydrates are DSH-I, $C_4A_3H_{13}$, $C_3AH_6$, CH, etc., and the hydrous phases are $C_2S$, $C_3A$, etc.

**Many traditional ceramists include (common) glass as a noncrystalline subset of ceramics.

ceramics in technological use are made at a high temperature and are thermodynamically stable at the temperature of preparation. Ceramics, of course, include an enormous range of compositions and structures, and hence properties. Cement and concrete are ceramics in this context. Technically speaking the term "mineral" means that the material occurs in nature. A term such as "artificial minerals" is not correct, strictu sensu--such materials should be called ceramics. Thus "alumina ceramics" is correct and adequate, instead of "artificial corundum," or "synthetic sapphire." The latter may, however, be valuable in describing the material to a non-technical audience.

(b) "Glasses," noncrystalline. Oxide glasses in this context, are distinguished unequivocally by two parameters, noncrystallinity and derivation from a liquid phase. Glass is not a specific material. Glasses include both hard candy and optical fibers. Properties of glasses vary over 10-20 orders of magnitude. Major misperceptions and misrepresentations have occurred due to failure to recognize this. Thus the citing of leach rates of tektites, obsidian, and $CaO-Al_2O_3-SiO_2$ glasses is scientifically invalid, if the goal is to present an estimate of the leaching of typical PNL zinc silicate or borosilicate glasses. All glasses are by definition metastable, but this raw fact in itself is not uniquely significant with respect to any particular glass's value in an SPI subsystem. The kinetics of reaction are also very significant parameters in all short-term considerations. In geological time (and obviously at higher temperatures where the kinetic barriers are more rapidly overcome) the metastability is significant. All glasses considered to date are insulator oxide glasses which are very roughly the same compositionally as the polycrystalline ceramics.

(c) Metallic. Some isolated metallic polycrystalline radiophases are formed within the target silicate glass radiophases of the reference waste forms. These consist principally of metallic Pd and Ru if the conditions are reducing. The only tailored metallic waste form is one conceptually designed by Macmillan et al. (19) and which is to be prepared by electrochemical reduction of halide melts from commercial waste streams.

Encapsulant Phases. Almost any "insoluble"* oxide or metal phase can serve as an encapsulant since the encapsulant is both a physical and chemical barrier. Obviously the best encapsulants will be the most "insoluble" phases. Again we can have the same three categories of materials.

(a) Ceramic encapsulants. $TiO_2$ (rutile), $Al_2O_3$ (corundum, sapphire), $3Al_2O_3$, $2SiO_2$ (mullite), $Fe_2O_3$, $MgAl_2O_4$ (spinel), are all outstanding candidates, with the anhydrous and hydrated calcium silicates of cement definitely in a "second class" group with respect to chemical resistance.

(b) Glass encapsulants. Very chemically-resistant glasses (at low temperatures) exist, and have been utilized for many years.

(c) Metallic encapsulants. Low melting temperature alloys based on Pb, have received the most attention, but copper and iron-nickel have also been studied in this role.

---

*The term "insoluble" of course, is relative. Moreover, it is highly dependent on the solvent, the temperature pH, Eh, etc.

The physical sizes and shapes of these encapsulant phases vary enormously, from a meter-long continuous (polycrystalline) metal phase surrounding glass beads to the more typical submicron cement, ceramic or metal phases. Details and references are provided in the next section.

## Processing

There are two major parts to ceramic (including glass) processing. The first involves getting or making the starting material, i.e. assembling the total *chemical* composition. The second involves the forming or consolidation (via one or more steps) into the final *physical* shape.

Particulate Forming and Mixing. Traditional glass and ceramic making processes have relied on the mixing of *dry solid powders*, or of mixing liquid + solid mixtures. If one considers that this involves mixing particles of, say, 50 μm (-325 mesh) and sintering by diffusion at 1000°C-1500°C (where cation diffusion rates may be in the range of $10^{-12}$ cm$^2$/sec), one realized how slowly this will occur. Clearly, mixing on the 1 μm scale, or better at the 5Å level *in solution* would provide 4-5 orders of magnitude gain in reaction speed. We thus have the following families of mixing processes available:

(a) Mixing solid powders. Inefficient, especially for solid state reactions when powders are not in micron range.

(b) Mixing in slurry with non-reactive liquid. Mixing better, but same objections exist.

(c) Mix in solutions (sometimes one component as sol) and convert to solid by

(1) evaporation [*spray-calcining*, also partly under (2) below];

(2) evaporation and decomposition simultaneously;

(3) gelation followed by desiccation followed by decomposition (sol-gel process).

Consolidation and Final Densification. One can group the well-known processes here in the traditional manner into three subsets defined by the starting material. This separates the processing options relevant to the radwaste problem into well-established traditional techniques in technology. Another critical parameter relevant to radwaste handling is the temperature of the process, since reliability and longevity of apparatus in remote operation will almost certainly be strongly dependent on temperature. We combine these two considerations in the following list of consolidation options available (Table 6).

## Canisters

These are typically meter-sized and made of metal--various steels, copper and titanium are the most prominently mentioned; but among the largest are those made of hot-pressed, theoretically-dense ceramics--three meters long and one-half meter in diameter.

There is no confusion on terminology here. The state-of-the-art will be reported on below for both metallic as well as ceramic containers.

TABLE 6

Consolidation Options for Ceramics

| TEMPERATURE RANGE | PHASE CHANGE | PROCESS |
|---|---|---|
| *High Temperature* (1000°C-1600°C) | Liquid → noncrystalline solid | Glass-making |
| | Liquid → noncrystalline phase, then to metastable solid, then to stable crystals | Glass-ceramic process |
| | Liquid → crystalline | Fusion-casting (controlled solidification) |
| (+ Pressure) | Fine-grained solids    larger solids | Sintering |
| | Solid phases reacted to give monolith | Hot-pressing (with or without matrix |
| *Moderate Temperature* (500°C-1000°C) | | |
| (+ Pressure) | Solid phases reacted to give monolith | Hot-pressing (with or without matrix) |
| (+ Pressure) | Solid phase + minor amounts of liquid, give monolith | Hot-isostatic pressing (with or without matrix) |
| *Low Temperature* (100°C-500°C) | Solid phases bonded into monolith | Cold (isostatic) pressing |
| | Solid phases bonded into monolith | Cold pressing (in metal matrix) |
| | Liquid phase surrounding large (cm size) units of solid | Molten metal encapsulation |
| *Room Temperature* + (electrochemical) | Solids mixed into slurry which sets; expel excess water | Cement (matrix) encapsulation |
| | Solids co-deposit with electrochemically reduced metal | Metal (matrix) encapsulation |

## Waste Loading, Characterization and Evaluation

"Waste Loading". Considerable confusion exists in the usage of the terms radionuclide content, fission product loading, and waste loading. Yet, this is a most important descriptor of any proposed waste form. It is one of the simplest variables to manipulate in order to achieve "better" properties. Thus, diluting a glass containing 30% fission products to one containing 7.5% in order to achieve a lowering of temperature in the canister is obvious. But in judging the efficacy of the two systems, it is essential that one compare the total system. The dilute system will require four times as many trucks for transportation, four times as much volume in the repository.

Moreover, there are three or more terms which are used to describe "waste loading." These are often used interchangeably, so that they often confuse even the expert reader. The question is: How does one express the concentration of wastes? As percent "waste," percent fission products, or percent radionuclides? Table 7(a) will show how different wastes have very different ratios for those three measures. Table 7(b) also tabulates the ratio of radionuclide and stable isotope for the commonest of the fission products.

This table tells us that there are enormous differences between various waste forms. The concentration of fission products plus actinides (excluding $U^{238}$) varies from about 1% and in the aged defense wastes at Savannah River, to ∿50% in some ceramic forms for commercial waste. The table also gives realistic estimates on the ratios between the total waste, the fission products, and the radionuclides. Thus the statement that a certain waste form can accommodate "30% of, say, a defense waste sludge" does not reveal that more than 99% of that 30% consists of nonradioactive inert oxides. Accommodating 30% of commercial wastes would require accommodation of 12% of radioactive fission products. It is clear that all descriptions of "waste loading" should be made much more carefully and explicitly.

Waste Form Characterization. The National Materials Advisory Board of the National Research Council defines characterization as follows:

"Characterization describes those features of the *composition* and *structure* (including defects) of a material that are significant for a particular preparation, study of properties, or use and suffice for the reproduction of the material." They go on to state:

"Most *property* measurements" do not "directly and unambiguously reflect the relevant compositional or structural features of a material," and cannot be accepted as valid characterization methods by themselves.

Up to the late seventies, much of the description or "characterization" of waste forms was done by measuring a single property-- leachability. Clearly this is not adequate. As the NMAB report recognizes, physical properties often provide simple sorting methods to avoid studying relevant samples in excessive detail. They are also needed to correlate with the characterization data. However, as in all materials characterization, there is a need to provide an appropriate (and balanced) level of detail in characterization. While in the past there has been no direct characterization at all, it is also easy to err on the side of excessive *and obscuring* detail.

TABLE 7a

Different Description of Waste Loadings Illustrating Why it is Essential to Distinguish Between Waste, Fission Product, and Radionuclide Loading

| Content or Loading of (wt % oxide) | Spent Fuel | Typical Commercial | | | | Typical Defense |
| | | (Battelle Glass) | | Ceramic | | (Encapsulant Ceramic*) |
| | | 72-68 | 76-88 | (SPC-4+U) | Synroc | |
|---|---|---|---|---|---|---|
| "Waste oxides" (calcines) | 100 | 25 | 33 | 80 | 10 | 95 |
| Fission product | 2.5 | 18 | 13 | 50 | 8 | <1 |
| An | 97.5 | 1 | 5 | 15 | 2 | 5 |
| Radionuclide | 98.5 (97.5+1) | 3 (2+1) ↑ ↑ Fp An | 7 (2+5) ↑ ↑ Fp An ($U^{238}$) | 25 (10+15) ↑ ↑ Fp $U^{238}$ | 3 (1+2) ↑ ↑ Fp ($U^{238}$) | ∿5 ($U^{238}$) |

*G.J. McCarthy, Candidate Crystalline Ceramics for Nuclear Defense Wastes, Nucl. Tech. 44, 451 (1979).


TABLE 7b

Relative Concentrations of Major Radioactive and Stable Nuclides in Fission Products

| Element | Radioisotopes (%) | Stable Isotopes (%) |
|---|---|---|
| Cs | 6 | 4 |
| Sr | 2 | 2 |
| Mo | – | 12 |
| Ru | 0.2 | 4 |
| Ce | 0.1 | 6 |
| Nd | – | 10 |
| Zr | 4 | 14 |
| Tc | 3 | – |
| Total | 15.3 | 52 |
| All Others | | 32.7 |
| Total | | 100% Fission Product |

With these caveats one can develop a set of normative characterizations as follows:

*(a) Microstructure.* Using optical, SEM*, and electron probe to identify the texture, morphology-porosity, relationships among phases, etc. Distinguishing between radiophases and encapsulant phases may be very significant.

*(b) Atomic or crystal structure(s).* In glassy or crystalline forms to identify by (S)TEM*, x-ray and electron diffraction the major phases (and any inhomogeneous regions in a glass), the minor phases, and the trace phases.

*(c) Bulk chemical composition.* (Including content of radionuclides, fission products and waste) and composition of individual phases to try to locate specific radionuclides.

The detailed characterization task in real waste forms is so overwhelmingly difficult and tedious, that the greatest care needs to be exercised in emphasizing *significant* details while omitting the rest. Much of the 76-68 glass from Battelle contains two or three crystalline phases in amounts up to 5%. The glass is phase-separated, on a 10μ scale. These factors may not affect its chemical resistance seriously. Typical tailored ceramics have 4-7 major phases detectable by x-ray diffraction and a similar number of minor and trace phases which need state-of-the-art STEM study. Whether it is worthwhile to determine each trace phase in a particular research mixture, when minor variations in waste stream may change these trace phases anyway, remains moot. What is called for here is a judicious mixture of using the relevant critical property measurement of "solubility" at repository 'p' and 't' conditions interactively with the compositional and structured characterization.

Waste Form Evaluation. For many years the standard IAEA leach test was the nearly universal performance test used for waste forms. Our recent research has shown that this leaching test is virtually useless and can be seriously misleading. Since the results of waste forms' reactions with water, *and with water and host rock under repository-simulating conditions* were first described (6) and discussed (4) by the Penn State group, this kind of hydrothermal testing has become one of the standard methods of evaluating a waste form. One of the major aspects of the present Penn State research is to develop a fundamental understanding of what is involved in such an evaluation. The topic of waste form evaluation forms the subject of another volume in this series and will not be treated in detail here.

---

*Scanning Electron Microscope = SEM; (Scanning) Transmission Electron Microscope = (S)TEM.

# Chapter 4. STATUS OF WASTE FORMS DEVELOPMENT: GLASS

## *Synopsis*

*A great deal of conceptual simplification is achieved once it is realized that there are only three major classes of waste forms: glass, ceramics and concrete-encapsulated. Within each of these classes, the compositional variations are minor and the processing is common. In other words identical processing equipment and steps are involved whether this glass is a borosilicate with 20% fission products or a phosphate glass with 15%; or whether an alumina-rich ceramic or a titania-rich one nicknamed "Synroc" is involved. Each of the three major classes is reviewed in turn starting with glass in this chapter.*

*Since the technology of radwaste-glass making has received perhaps 99% of all the funding to date, a great deal more is already known about it. The approach taken in this chapter is to set the status of the glass technology in perspective. We provide only in outline a description of the process of making the radwaste glasses. The wide-ranges of composition in use, and the many disparate processes proposed are brought together in tabular form, showing that glass-waste forms are far from identical to each other.*

*In order to describe the properties achieved with glassy waste forms, we can turn instead of to research samples, to full-scale simulated production samples. Since an excellent report on such "production" waste glasses by J.E. Mendel is already extant, we do not treat it separately here but insert the entire article as an appendix.*

# STATUS OF WASTE FORMS DEVELOPMENT: GLASS

In this section, an attempt is made to provide a review of the status of the scientific and technological basis for each of the important waste forms being considered for solidification of high level wastes. As has been noted before, at the present time the non-specialist needs a clear overall picture of how each waste form works, how it is related to others, the stage of development, the R&D backup which exists, and the advantages and disadvantages of each. Especially since there is a great unevenness in development, and even more so in knowledge about different waste forms, we will attempt to be tutorial in approach, spending more time on the lesser known forms. These sections on waste forms will be followed by a set of references, and then a bibliography relevant to the waste forms, taken from the most important conferences of the last two years. Thus, the summaries about each waste form which follow are those judged to be most valuable to the audience we expect will most likely be interested in this material. The waste forms we will treat are various combinations of these radiophase and encapsulant phases:

Radiophases

- Glass (mainly)
- Ceramics (many kinds)
- (• Metals)

Encapsulant Phases

- Metals
- Ceramics
- Concrete
- Glass

## Glass as the (Nominal) Radiophase

Overview

In the brief historical overview of chapter 2, it was indicated that glass has received the lion's share of development effort as the radiophase of choice. The technical origin of this choice is obscure. The oversimple concept of glass as a universal solvent had much to do with it. High temperature silicate (or phosphate) melts do indeed dissolve various amounts of most other oxides. But the range of maximum solubility is enormously variable. Moreover the range of compositions which form bulk glasses is a miniscule fraction of the universe of oxide compositions. Thus there was little comparative research to back up this choice at the time when the choice was made. Such *research* is now being conducted worldwide in a desultory manner, while the *development* of glass-making as the "reference process" proceeds, especially in France.

A great deal of literature exists on the reference process for making glass as a radiophase. The *process* consists of melting selected compositions of waste together with additives in the form of a fit and casting into a monolithic glass block inside a cylindrical steel container or, in a somewhat more advanced version, forming beads for encapsulation into a metal matrix. And we include as the next section a detailed report on the properties of the *products* made.

The process can be conceptualized in simplified form as follows to show the choices:

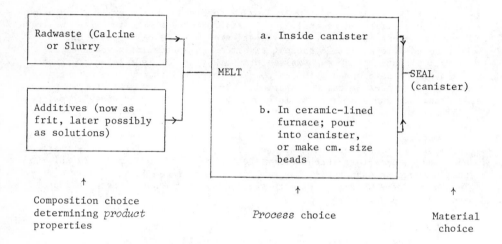

Figures VI(a) and (b) and VII show schematically for both in in-can and ceramic melt options how these steps have actually been translated into reality at pilot plants at various locations all over the world, and in a working hot plant at La Hague in France.

If one examines the literature regarding an evaluation of the "product" and the "process" choices in making glassy waste forms, one can summarize the relative merits as follows:

Process
- considerable R&D backup, and French hot experience
- good process control—mixing, melting, pouring
- relative simplicity.

Product
- good resistance to leaching in *low temperature* screening tests
- has shown no major deterioration with simulated radiation exposure ($\alpha$, $\beta$, and $\gamma$) equivalent to total dose over $\sim 10^6$ years from actual radio-content.

The limitations cited in the literature are:

Process
- It is a very high temperature process for long-term remote operation
- Experience advantage dubious, especially with probably stretchout in U.S. timetable, and compared to track record on ceramics and calcine.

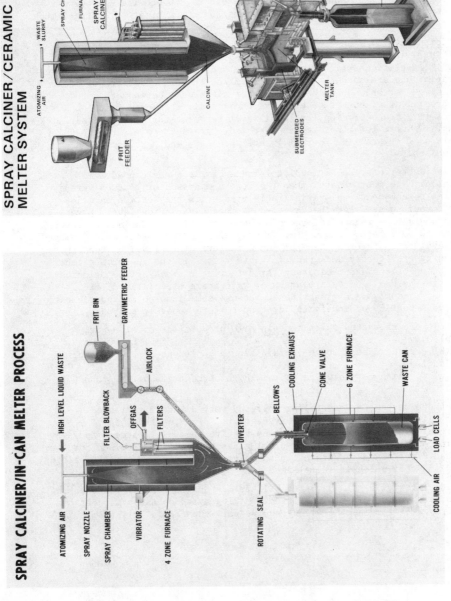

Fig. VI (a) Spray Calciner/In-Can Melter, Schematic.
         (b) Spray Calciner/Ceramic Melter, Schematic.

$$\text{Product} \left\{ \begin{array}{l} \text{- Requires that temperature must be kept low, or} \\ \text{repositories dry, even if this means dilution and} \\ \text{hence increase in volume of waste} \\ \text{- Is not as adaptable to waste composition change as} \\ \text{imagined.} \end{array} \right.$$

The summaries (7c) referred to are reasonably up to date on glass and we can add little to them. Here we would like to add some additional perspectives from the materials and geosciences to stimulate the reader to place glass into a framework appropriate to balanced comparisons among waste forms.

The following are some less widely appreciated features of the vitrification process and its resultant products, to provide a framework for comparison with the newer alternatives about which, at this stage, much less is known.

Compositions. The compositions of glasses developed and used *prominently* in rad-waste research vary enormously. Table 8 lists some typical compositions which have been developed in different laboratories. The variations are such that to a glass maker these are rather different materials. Not only do we have the major difference between the phosphates and borosilicates, but we observe the anomous differences in fission product (FP) concentration, and similar differences for $Al_2O_3$ and alkaline earths. We note for completeness that many "glasses" (e.g., PNL 76-68) contain some crystalline (ceramic) material--$RuO_2$ and *metallic* Pd as radiophases.

The matter of compositional tolerance of glass has caused some confusion. The most general theory of the comparison of glasses versus ceramics to compositional variation is treated in the section entitled Ceramic Radiophases. At the day-to-day process variation level, a well-designed glass is indeed tolerant to compositional fluctuations of several percent in most major components. All candidate RWM glasses have achieved this level of engineering design. At an intermediate level of concern between general theory and specific waste glass, is the question of the utilization of a glass radiophase for different classes of the principal wastes in the U.S. Certain classes of wastes are especially difficult to obtain *efficiently* in a glassy host, because some ions do not participate in network-forming or modifying but tend to form separate phases:

(a) Defense wastes which contain high proportions of $Al_2O_3$ or $(Al,Fe)_2O_3$.
(b) Defense wastes which contain high proportions of $ZrO_2$ and/or fluoride ions.
(c) Civilian wastes which contain high percentages of $ZrO_2$ and $UO_2$ (TMI).

Among these wastes (a) and (b) are already very dilute with respect to radionuclides. Yet they require even greater dilution than, say, commercial PW-4 compositions. Or they require special treatment to remove certain constituents. Thus, a typical glass composition proposed for the Savannah River sludges contains in the final glass only 28% of the original *sludge* (most of which is inert), and only $\sim 0.1\%$ of fission products.

The Three Mile Island wastes will form a unique case if reprocessing of the damaged fuel is not permitted. Making a radiophase glass out of this $ZrO_2$ rich composition would require very substantial dilution.

TABLE 8

The Wide Range of RWM Glass Compositions in Use

| | % F.P. (± actinide) | | Approximate Composition (as oxides wt %) | | | | | |
| | FP | An | Alkalies | Alkaline Earths | $Al_2O_3$ | $SiO_2$ | $B_2O_3$ | $P_2O_5$ |
|---|---|---|---|---|---|---|---|---|
| Nepheline-syenite glass (Canadian, 1960, Ref. 20) | <<1 | <<1 | 13 | 15 | 20 | 51 | | |
| Fingal (U.K., ~1966, Ref. 21) | 11 | ~3 | 12 | | 5 | 43 | 13 | |
| PNL-72-68 (U.S., 1976; Ref. 7c) | 24 | 1 | 8 | 6 | | 27 | 11 | |
| PNL-76-68 (U.S., 1978; Ref. 7c) | 13 | 5 | 13 | 3 | – | 40 | 10 | 0.5 |
| Marcoule (Ref. 22) | 13 4 | ? | 14 17 | 1 2 | 8 13 | 42 40 | 18 16 | |
| USSR experimental (1979; Ref. 23) | 20 25 | ? | ~20 ~25 | ~5 | | ~35 | ~15 | ~50 |
| Belgian vitromet (1978) phosphate (Ref. 24a,b) | 28 | →0 | | | 5 | | | 50 |
| SRP (proposed) (Ref. 25a,b) | 0.1 | 1 | 22 | 3 | 3 | 40 | 7 | |

## Batch Mixing in Glass Preparation

In Table 9 we summarize some of the process options in glass manufacture. The first column lists the different approaches to the first step in any glass or ceramic process: the mixing of the batch or raw materials to yield the final composition. The traditional techniques of mixing glass batches are:

(1) Mixing the *solid* fine radiophase powders from the calcine with solid glass frit. The different styles of calciner provide minor process variations.

(2) Mixing the radwaste *liquid* stream with a pre-melted frit or a slurry of glass-making components and making a super- or tailored-calcine of the final glass composition.

## TABLE 9

### Process Options in Glass Making

| Type of RWM Glass | Batch mixing for glasses | Melting |
|---|---|---|
| U.S. Reference | HLW as solid calcine stream + solid glass frit | (a) In-can melting (up to present)<br>(b) Joule-heated small tank (1978→) |
| Harvest, Fingal U.K. Reference | Glass making components as slurry + HLW as liquid stream | In-can melting (1965-1979) |
| Marcoule | Rotary calciner + frit | Metallic melter |
| Mol-Jülich | Denitrated and special calciner to yield aggregates (1 mm)<br><br>HLW liquid + slurry calcined | Electrically heated (current) platinum bucket to make beads<br><br>In-can melting |
| Possible future | HLW + all glass-making components mixed *as liquids*, then spray calcined, or desiccated after gelation | Any of the above |

*General Reference:  Ceramics in Nuclear Waste Management, Ed. T.D. Chikalla and J.E. Mendel, CONF-790420, Department of Energy publication (1979).

These techniques form the backbone of the present glass waste form preparation process.  There is one new development on the horizon:  this is the use *here* of liquid-phase mixing followed by calcining or gelation and desiccation.  It has been applied extensively for batch preparation as crystals or glasses in this laboratory for nearly 30 years.  It was applied explicitly to bulk glass-making by McCarthy and Roy (26a), and has since been followed up by many others (26b,c).  It assumes the greatest homogeneity of feed for a melter, and minimum melting time and temperature. Further details on the process are provided under the ceramics section where this is even more relevant.  It is one of the new technologies to monitor in relation to glass processing.  The very recent work by Pope and

Harrison (26d,e) have shown the substantial advantages of the gel route in permitting access to high alumina glasses, and of making glasses at lower temperatures.

## Melting Processes

Table 9 also shows the three basic classes of glass-melting technologies in use. The first is the "in-can melter" or external heating of the alloy canister, the second a small tank furnace heated electrically by Joule heating of the melt itself using molybdenum or Inconel electrodes, and the third is external heating of metallic tanks. The three processes have all been operated "hot" in small batches for short times. Remote continuous hot operation has been started up in the Marcoule plant and production and reliability data are being gathered in order to assess the efficiency of the technology. Successful research operation of U.S. melting and pouring operations has led to some full scale canisters of hot solidified glassy waste. The U.K. Fingal process also produced several canisters of hot glass many years ago, by in-can melting, but full scale production at Windscale is not expected for some years. *The trend at present in the U.S. and Germany following France is toward the small tank, but it is evident that the advantages of one over the other are not very large since no clear decision has been made by all concerned for one or the other.*

## Properties

Extensive work has been done on the chemical (principally "leaching") and physical (thermal conductivity and frangibility upon impact) properties of candidate glasses. What has been established (for details see later) is that at room temperature leaching rates in the range of $1 \times 10^{-6}$ gm per day per sq.cm. are attainable. At 100°C, these numbers increase by an order of magnitude. This degree of general "insolubility," equivalent roughly to that of window glass at near-ambient temperatures and near-neutral pH's gave the glass radiophases their viability as the candidate reference form.

It was only when it was shown (6) that glasses were very rapidly altered at higher temperatures (200°C and above) with water present, that the upper temperature limitations on the use of the glass radiophase were realized. It has also become clear that the waste form-water-host rock interactions drastically influence "leachability" or rate of attack. Brine solutions attack glass more slowly than solutions with other rock types. A wide range of experiments are now in progress to obtain kinetics of weathering at the intermediate temperatures. This is clearly one principal frontier of glass RWM research: determining rates of attack as a function of temperature and ambient and composition, especially waste loading. Judgments can be made on the basis of these studies regarding the maximum temperature specific glasses can be exposed to and their interaction within the total subsystem.

The thermal conductivity and mechanical properties are roughly in the same range for all probable glasses. Radiation damage studies at many laboratories, especially using $^{244}$Cm doping, have established that only minor (1-2%) volume changes and little change in leaching behavior occur due to $\alpha$-damage (28). There have apparently been no direct studies of *transmutation* induced structural effects in glasses. So far, transmutation has been considered within the overall effects of radiation in fully radioactive glasses.

While many of these properties refer to laboratory scale experiments, the interest in glass as a waste form has moved to an engineering scale and in the next section we present a summary of the properties of fully engineered products.

# GLASS AS A FULLY ENGINEERED WASTE FORM

While the actual laboratory-scale scientific data on glass as a radiophase is perhaps not much more extreme than any of the alternatives, there has been a much larger accumulation of data and experience with large-scale engineering on glass as a radiophase. Indeed extensive data are available on the properties of engineered waste glasses. This section provides a summary of such data, including data on hot (radioactive) specimens as well as full-scale 'cold' engineering samples. This section is reproduced verbatim from a Report entitled 'The Storage and disposal of Rad-Waste as glass in canisters,', PNL 2764, UC-70, by John E. Mendel of the Battelle Pacific Northwest Laboratory, by whose permission it is reproduced here; it is presented separately as Appendix 1.

# Chapter 5.  ALTERNATIVE WASTE FORMS:
## CERAMIC RADIOPHASES

### *Synopsis*

*Ceramic waste forms were the earliest forms considered for solid phase immobilization in the fifties, and actually produced. After a long hiatus interest in them was revived in the seventies and they now constitute the major alternative to glass with the especial advantage of greater stability in the presence of water at moderate temperatures.*

*Although the term "ceramics" often includes (non-crystalline) glass it is used here to designate crystalline materials. All non-metallic inorganic crystalline radiophases are accurately and generically designated by the term ceramic. Brand names and nicknames such as "Stopper" and "Synroc" suggest misleading differences where none exist. Specific ceramics are properly distinguished by the well-established parameters of chemical and phase composition and microstructure.*

*The concept of using those minerals, which have been shown by petrologists to survive the pressure and temperature conditions similar to those in a projected waste repository, as models for ceramics is described. Using this strategy and crystal chemical principles, the tailormaking of a set of 4-8 separate crystalline phases in a ceramic was achieved in the mid seventies. A slight variant of this strategy was to use only titania rich phases, thereby being restricted to much lower levels of fission product loading. The concept of the desirability of hydroxylated ceramics because they are stable under the more recently proposed temperature conditions (100-200°C) for repositories is included.*

*The status of the processing of ceramics--both powder preparation and consolidation--is described in some detail since it is unfamiliar to any. Finally special cases of ceramic radiophases--ranging from spent fuel to the doubly-coated ceramic pellets embedded in metal are described.*

# DEFINITIONS, THE MINERAL-MODEL CONCEPT

Ceramics are often described as inorganic non-metallic materials. Common examples are the bricks, cups and saucers and bathroom fixtures in households: less well known are the capacitor materials, the lighting and TV phosphors, the SiC and $Al_2O_3$ in abrasive and cutting tools and all the high temperature refractories in the metal-producing industries. The crust of the earth is, of course, made up largely of precisely these same materials. If they are natural they are called minerals (for single phases, e.g., *feldspar*) or rocks (typically multiphasic, e.g., *granite* = quartz + feldspar + mica). All "artificial mineral" assemblages are by proper usage called ceramics.

To scientists trained in the geological sciences, it is virtually axiomatic that if one were to find a solid host for radio-ions which would have to retain its integrity under conditions of what is called low-rank metamorphism, one would turn to those minerals which have proven themselves to have survived for hundreds of millions of years during and after exposure to these conditions. Moreover, since the late forties, the science of experimental petrology has made it possible to simulate geological events rather well the better the higher the temperature. Thus an enormous body of experience and expertise exists on predicting the stability or interreaction between phase assemblages under conditions of elevated temperature and pressure.

It was on these grounds as noted in Chapter 2 that McCarthy and Roy developed the generic concept of using the resistate minerals as structural models and applied it to the making of ceramic assemblages to accommodate the typical compositions of a reprocessed commercial waste stream.

## Ceramic Radiophases

We shall treat the entire family of ceramic materials in the same sequence as above for glasses.

Use of Brand Names: "Ceramic Sponge," "Supercalcine," "Stopper," "Synroc". The practice of putting "brand names" on processes and products has a long, if not honorable, history in radioactive waste management. Vitrification processes and products have been given names: FINGAL, VERA, HARVEST, PAMELA..... Ceramic processes and products have been called CERAMIC SPONGE, STOPPER..... [see Isaacson (10)]. The term "supercalcine" was applied by McCarthy (29), to the *process* of *chemically mixing all components to yield a predetermined total phase composition* which would be refractory and highly insoluble. "Super*calcine*" itself is exactly what it says--the fine precursor powder as a *process* step towards the final tailored ceramic. Supercalcine-*ceramics*, which implies the tailored ceramics based on mineral models, are the *product*. Although in all his early work McCarthy (11) used the generic term "ceramic waste forms," the term "supercalcine" continues to be *erroneously* used by many as though it referred to a product, indeed a particular composition. Supercalcine can refer to *any* composition. McCarthy's many early sets contained oxides, silicates, phosphates, and tantalates (see Table 10). Ringwood named his

TABLE 10

The Sets of Phase Compositions of Various Ceramic Waste Forms

(a) For commercial wastes

(b) For Savannah River defense waste

Illustrating the Essential Similarity of Different Approaches

to Designing Ceramic Waste Forms

**(a) COMMERCIAL PW-4**

| High Concentration of F.P. (McCarthy et al.) | | | Low Concentration F.P. (Ringwood et al.) |
|---|---|---|---|
| (i) | (ii) | (iii) | |
| Scheelite | Scheelite | Perovskite | Zirconolite |
| Fluorite | Fluorite | Fluorite | Perovskite |
| Pollucite | Pollucite | Pollucite | Hollandite |
| Apatite | Zirconia | Monazite | ( + ?? ) |
| ( + ? ) | Monazite | Spinel | |

**(b) DEFENSE**

| McCarthy (Conceptual) (1979) | PSU/Rockwell (1981) | Rockwell/PSU High Al         High Fe (1981) | | "Synroc D" (1981) |
|---|---|---|---|---|
| Spinel | Spinel | Spinel | | Spinel |
| Nepheline | Nepheline | Nepheline | | Nepheline |
| Uraninite | Uraninite | Uraninite | | Zirconolite (≃Uraninite) |
| Corundum | | Corundum | Perovskite | Perovskite |
| Brownmillerite | Wustite | − Magnetoplumbite − | | Hollandite |
| | + ? | | | + ? + ? |

first *dilute* ceramic waste form SYNROC (synthetic rock), but has since been developing new sets of phases, and labelling each set A, B, C, D, etc. *Many have thus been led to think that Synroc is an alternative to, instead*

*of being but one more specific example of, tailored ceramics.* We observe
in Table 10 that one need only add to the list of sets of phases proposed
by McCarthy for ceramic waste forms, either or both hollandite and
zirconolite (an ordered fluorite), to arrive at the "typical" Synroc
formulations. From the following examples:

Synroc A:  "Hollandite" + Perovskite + Zirconolite + Ba-Feldspar +
Kalsilite + Leucite.

Synroc D:  Hercynite (= spinel) + "Hollandite" + Perovskite +
Zirconolite (∿Fluoride + Nepheline (30).

"Supercalcine-ceramic:"  Perovskite + Scheelite + Fluorite + Pollucite
(12),

we will find that one cannot pick out any subset by any logical criterion.
Table 10 makes the point in greater detail that 'Synroc' is simply one
other ceramic, and in the case of Synroc-D, it is essentially identical to
one of McCarthy's ceramics. It is obvious that these sets are in fact
individual attempts to *tailor ceramic radiophases* as waste forms. Regret-
tably, great confusion has resulted from the use of such brand names,
especially for the large technical, but non-specialist, audience. We
strongly recommend that this practice be discontinued forthwith. "Tailored
ceramics" or just ceramic waste form is a sufficiently generic and descrip-
tive term that we have been using, and will continue to use, to describe
all crystalline ceramic products. Every specific tailored ceramic *product*
must be described by its total composition, set of phases (structure types)
and microstructure. In addition, the *process* used to make the product
should also be specifically described (see section on characterization).
In contrast to the above-mentioned confusion, the practice of the Battelle
PNL group on assigning specific numbers (e.g., 76-68) to specific compositions is commendable. In both cases, of course, the composition by itself
is only a small part of the description needed.

## Composition of Tailored Ceramics

Following the principle described in the introduction of tailormaking
using a combination of mineral-resistates as models, McCarthy and
co-workers examined several dozen combinations of simple oxides, ternary
titanates, zirconates, silicates, phosphates, etc., over the years 1971-79.
The systematic approach to the problem *for high concentration wastes*, which
also required the phases in the assemblage to be mutually compatible at
high temperatures, started by reacting the total HLW oxides composition,
identifying the phases, and making additions as needed to change some of
the less desirable phases towards mineral-resistate phases (11b,29).

In the early to mid-1970's, one of the challenges undertaken by the
Penn State group was to design assemblages of *compatible* crystalline
radiophases that would simultaneously provide reasonably *high loadings* of
radionuclides (comparable at least to then current borosilicate glasses),
minimal volatilization during processing, low water solubility, and that
would be able to be processed with simple, remotable, and preferably
already available, equipment. When these boundary conditions are super-
imposed simultaneously--and they are indeed essential in an optimum
system--the choice of radiophases is narrowed considerably. To cite one
example, while many minerals and synthetic mixed oxides are capable of

incorporating 0.1-1 wt % of Cs into their structures, only a few phases incorporating major Cs meet the above-listed criteria: $CsAlSiO_4$, $CsAlSi_2O_6$ (pollucite), and $CsAlSi_5O_{12}$ with ∿54, ∿45, and ∿26 wt % of Cs content, respectively. The structure types of other high-loading radiophases studied by McCarthy and co-workers (31a,b) with the respective radionuclides contained include: monazite (lanthanides [Ln], actinides [An]), feldspar (Sr), powellite (Sr,Tc), rutile (Tc,Ru), apatite (Sr,Ln, An), perovskite (Sr,Ln,An,Tc,Ru), and fluorite (Ln,An). *Additional* structure types that would contain principally the large amounts of "cold" ions always found in HLW (e.g., Mo,Fe,Cr,P,Ni,Na,Te,Cd,Ag) include the above plus baddeleyite, nepheline, spinel, corundum, sodalite. Ringwood added hollandite, and the Rockwell group, magnetoplumbite as important candidates; and most recently we added $CsZr_2P_3O_{12}$ in the NZP structure.

The theory is that any *oxide mix containing a high concentration* of radio-ions would likely involve several of *these* phases. *Dilution by use of nonradioactive ions of the same species was always obviously available* as a strategy to retain the desirable stability characteristics, while reducing the thermal loading, radiation damage, and transmutation effects. As in all dilution strategies, whether in glass or crystals, this would have significant costs in terms of volume of waste, but it needs absolutely no additional research other than to define how much cold isotope dilution would be necessary for the desired reduction in temperature and/or radiation effects.

### a. Typical Compositions Used to Date

Table 11 lists the principal ceramic phases which have been proposed as suitable for the incorporation of substantial amounts of radionuclides. It is obvious from the inspection of this table that there is an enormous number of combinations of phases which can incorporate all the wastes in relatively insoluble combinations. Table 10 has already listed some of the sets which have been utilized. Many more are possible and likely. However, there are three crystal chemical equations to be solved simultaneously. One is to match the availability of coordination polyhedra in the structures selected, to the composition of the waste, and to be certain that the resulting phases are compatible (i.e., actually form) under the processing conditions selected. The composition of Purex waste displayed in terms of ionic radius and concentration is shown in Fig. VII. This determines *both the phases and their ratio* which must be used to incorporate all the radionuclides. This is a non-trivial exercise in molecular engineering when one considers that one has about forty elements *in a fixed ratio* in a typical waste stream. McCarthy's extensive data suggests that for high concentration radiophases, a half-dozen major phases will be necessary, probably with two or three additional trace phases. Ringwood's set (B) (now C) is misleadingly simplified, since in fact even at the much lower dilutions he uses, at least three or four *other* phases will form. We will show later that the presence of these minor phases may be of little consequence under certain conditions. In any case, two other boundary conditions must also be met. The phases will need to be close to thermodynamic equilibrium with each other, since they may be made at high temperatures. Next, they should be close to equilibrium with the host rock. The work of McCarthy and his students in solving these simultaneous equations led to the choice of several preferred sets over the years.

Table 11

Most Important Candidate Ceramic Radiophases

| Structure Type** | Illustrative Compositions | Possible Radionuclides and other F.P. incorporated | | |
| --- | --- | --- | --- | --- |
| | | Total | Major | Minor |
| [Monazite] | $(Ln,An^{3+})PO_4$ $An^{4+}SiO_4$ $(Ca_{0.5}An_{0.5}^{4+})PO_4$ | Ln,An | | |
| [Perovskite] | $(Sr,Ba)(Ti,Zr)O_3$ | Sr | $Ru,Tc^{4+}$ | Ti,Cs,An,Ln |
| [Pollucite] | $CsAlSi_2O_6$ | Cs | | |
| [Spinel] | $(Ni,Fe)(Al,Fe)_2O_4$ | | $Tc^{4+}$ | Ru |
| [Apatite] | $(Ca,Sr)_5(PO_4)_3$ | Sn,Ln | | Ru |
| [Scheelite] | $(Sr,Ba)MoO_4$ | Sr | $Tc^{6+}$ | Ln,An,Ru,Te |
| [Feldspar] | $SrAl_2Si_2O_8$ | Sr | | |
| [Fluorite] | $(An,Ln,Zr)O_{2+x}$ | Ln,An | | Sr |
| [Zirconolite*] (ordered fluorite) | $CaZrTi_2O_7$ | Sr | Ru | |
| [Rutile] | $RuO_2$ | Ru,Te | | |
| [Fcc Metal] | Pd | | | Pd |
| [Hollandite*] | $BaAl_2Ti_6O_{16}$ | | Cs | |
| [Nepheline] | $NaAlSiO_4$ | | | Cs(?) |
| [Barium-aluminate* iron titanate] | $Ba(AlFe)_2Fe_8Ti_{13}O_{38}$ | | Cs | |
| [Magnetoplumbite*] | $BaFe_{12}O_{19}$ | Sr | Cs | |
| [Na-Zirconophosphate = NZP] # | $NaZr_2P_3O_{12}$ | Cs,Sr | Ln,An, $Tc^{4+}$, Ru,Te | |

*Cs solubility in BAIT and hollandite, first proposed by Ringwood, and in magnetoplumbite by the Rockwell Science Center group, has not yet been quantitatively determined.

**Square brackets in crystal chemical convention indicate that it refers to the structure-type and not the composition.

#Separate substitutions achieved by Alamo, Vance and Roy (Mat. Rès. Bull., in press).

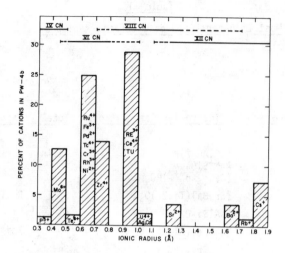

*Fig. VII. Composition of commercial waste displayed as concentration of
ions of different size. This is the parameter which controls the ceramic
radiophase structures needed to accommodate the ions.*

Table 10 has listed the several 'best' sets of ceramic radiophases which
can accommodate the fission products *in the as-received ratios in commercial waste*. It also lists one set of host phases rich in titania for
dilute accommodation of commercial wastes. Surprisingly, in most of these
titania ceramics the percentage of the normal *fission product mixture* which
can be accommodated, has not been determined to date.

### The Dilute-Ceramic Crystalline Solution Strategy

In the glass case, the fission product concentration can be, and has
been, varied from nearly 1% to nearly 20%, merely by adding more frit. In
the crystalline case, dilution is obviously also possible merely by adding
more nonradioactive ions of the same species as are present in the fission
products, to give the same structures. However, in the crystalline case,
dilution has another crystal chemical advantage. Many more compatible
phase assemblages exist, made up of structures that will accept a small,
say 1%, level of incorporation of an ion, than will accept a structure-
constitutive role for the same ion (say, 20-50%). Thus, $Sr^{2+}$ can no doubt
be accommodated in many spinel structures up to, say, a 1% level, whereas
there are no Sr-spinel oxide phases.

This *dilution* concept was the modification of the McCarthy-Roy
tailored ceramics introduced by Ringwood. He argued that lowering the
radionuclide concentrations greatly from those of "supercalcine-ceramics"
would open up numerous new "mineralogies" (i.e., ceramic phase assemblages).

*"Accordingly, we will now describe a more expensive, but relatively simple process based upon immobilisation of radwaste in crystalline phases, in which the concentration of radwaste does not exceed 10 per cent. Moreover, if considered desirable, this percentage could be further reduced to about 2 per cent in a second stage of processing which would take place about 10 to 20 years after the first stage....*

*However, if the added components exceed 90 per cent, they will control the nature of the crystalline phases produced. The radwaste atomic species will then simply substitute in low concentrations within the crystal lattices determined by the major added components." (emphasis added, from Ringwood, Ref. 13a)*

Thus, Synroc A and B (now C) designed for commercial hot liquid waste had waste oxide loadings of ∿10%*, compared to ∿80% for the then current "supercalcine-ceramics." He argued that an increase by about an order of magnitude in waste and repository volume would have little impact on the overall costs of radioactive waste management. While this may not be a question for the waste form designer to decide, what is of interest in the present context is whether in fact any major advantages in properties can be achieved with the less restrictive crystal chemical demands, to offset the volume disadvantage.

The principles governing the extent of crystalline solubility are well known *qualitatively*. The most important parameter determining whether a particular ion will substitute for another is the difference in ionic radius (r). Experimentally, it has been shown that the latitude or tolerance for $\Delta r$, is of the same order (for substitutions at the few percent level) that also corresponds to changes of coordination. These facts have suggested that as few as *two* structures are needed to provide XII, VIII, VI, and IV coordinated sites for all the ions in radwaste. Hence, any combination such as perovskite (XII and VI) + apatite (VIII + IV); or garnet (VIII + VI + IV) + mica (XII + VI + IV), etc., would be able to accommodate all the fission product radionuclides at a few percent level.

A not well-appreciated corollary to this dilution strategy is that, in an unpartitioned waste, the ultimate loading is limited by the solubility of the hazardous element *least* soluble in the design phase assemblage. For example, Cs has a more-or-less fixed proportion in reprocessing wastes. Thus, if Cs had the least crystalline solubility in a particular set of phases, the total waste loading achievable in the assemblages will be limited by the Cs-solubility in the chosen Cs fixation phase. It is meaningless to state that the loading factor for some single waste components is 30% or 60% or even 100% because that component will always have a concentration proportional to Cs and can only be loaded up to a value proportional to the Cs loading. By the same argument, it is also certain that structural sites must be provided for all the ions in the

---

*In fact there were no data to establish even the solubility of this level of PW-4 or PW-7 waste loading. Probably no more than 2-3 percent of commercial waste oxides can be accommodated in the three phases of proto-typical Synroc B.

waste. It was, therefore, obvious from the outset that only the three cited phases of Ringwood's ceramic-B could not accommodate the typical fission product spectrum, since none of the phases hollandite, perovskite and zirconolite had fourfold sites for the smallest cations such as the $P^{5+}$, $Si^{4+}$, and $Mo^{6+}$ (in air) which are ubiquituous in wastes.

b. Low Temperature Ceramic Radiophases. We have noted above that much of the early work using clay and zeolite templates did in fact claim to adsorb all the radionuclides from waste streams. Some early research was devoted to converting these into low temperature ceramics with layer-structure phases. Schulz and Kupfer (32) and Barney (33) reacted kaolinite and bentonite with alkaline waste solutions at 100°C, producing mainly the mineral cancrinite. Presumably all other phases were noncrystalline. Some leach rates as low as $10^{-7}$ to $10^{-9}$ gm/cm$^2$ day below 100°C were reported. Further incorporation of these cancrinite radiophases into calcium silicate matrices was also noted. However, these promising low temperature approaches seem to have been abandoned after very little work. Roy (34) has emphasized that from a thermodynamic and mineralogical viewpoint it is certain that hydroxylated phases such as mica apatite, nephelene hydrate or clays will be more stable and hence even less soluble at temperatures in the 100°-200°C range in the presence of water than the anhydrous ceramics.

## Relative Adaptability to Compositional Change of Ceramics and Glass

A major misconception exists regarding this issue. There is an intuitive sense that just as with an aqueous solution, a glassy solution can accept a very wide range of solute concentrations. This is of course, true up to a point, although quite obviously the properties of the glass will change more or less continuously as a function of composition. What is apparently much less appreciated by non-specialists is the basic fact that glasses in general are a very tiny and special class of solids. Even making a true (single phase) glass in large blocks is very demanding on compositional ranges. (Witness the fact that the standard U.S. reference Battelle "glass" has about 5% of [two] crystalline phases!)

However, with respect to ceramic waste forms many serious erroneous ideas and statements exist with regard to "compositional forgiveness." These statements betray a basic failure to understand the phase rule. A simplified example will explain the principle. Most common ceramics are polyphasic: a porcelain for instance may contain feldspar, mullite and a $SiO_2$-phase. What happens if we change the CaO content or the $SiO_2$ content? To answer that we need the relevant phase diagram (just as we need the much more qualitative assessment of the so-called glass-forming region in a particular system). If the phase diagram shows that phases A, B, and C are stable together then what it really says is that within the compositional triangle, the ceramic is "infinitely forgiving." i.e., Any starting composition within the triangle A-B-C will give only the phases A+B+C at equilibrium. Since all the properties of the ceramic are determined by additivity rules from those of the constituent phases, these properties will vary smoothly from any part of the triangle to any other. Thus in Fig. VIIIa a ceramic with a normal composition X, can tolerate 33-66 mole % fluctuations in any direction and contain exactly the same phases.

In both glassy and crystalline systems one can, of course, find compositions which are more or less forgiving than others, as is illustrated in Fig. VIIIb. In the glassy case, some compositional changes will cause precipitation of crystalline phases: in the ceramic case new phase sets will form. Both of these *could* have less desirable properties.

This treatment of the "forgiving" nature of ceramic systems is the most elementary one. In real life one finds that each phase can itself be a crystalline solution (exactly analogous to glass). In practice, then, a ceramic is a mixture of two, three, or more phases, each of which is a crystalline solution. It is here where another part of the misconception arises. The typical crystalline solution taken as a single phase is rather strict in its demands on the chemistry it can accommodate. The compositional fluctuations permitted within such a phase are often binary even in a 3- or 4- or 5-component system. This, however, in no way detracts from the forgiving nature of the multiphase *assemblage* (ceramic) according to the principles outlined above.

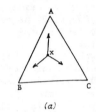

*(a)*

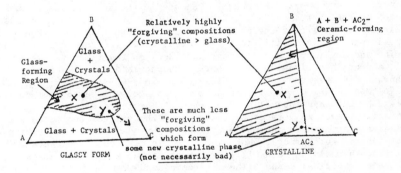

*(b)*

Fig. VIII.  *Comparison of relative compositional "forgiveness" in glass and ceramic radiophases.*

(a)  *In a crystalline assemblage, ABC, no compositional change within the compatibility triangle can change the phases (only the relative amounts of the phases will change).*

(b)  *Note two target compositions (x) and (y). It can be seen that in compositions such as x, the crystalline assemblage can be much more forgiving to change of total composition than the glass. It accomplishes this without change of phase. On the other hand, other optimized compositions, such as y, may be much less forgiving in both glassy and crystalline forms. In both cases small additions of C form a new phase.*

## Processing: I, Mixing (Powder Preparation) of Ceramic Compositions

The recent results on waste form stability show that the difference in product characteristics may well be overshadowed by differences in process simplicity and reliability in remote operations. Thus innovation in *processing* may pay even greater dividends in alternatives to traditional approaches.

The preparation of ceramic batch compositions faces analogous problems to glass batch preparation. In both cases, inhomogeneities require extra reaction time at high temperature. In the ceramic case with mainly solid state reaction, these times may be very long indeed, and hence there is a premium in preparing very intimate mixtures of very fine powders. An alternative strategy for making homogeneous ceramic powders in multicomponent systems has emerged over the years. Korosy (35a) first discussed the simultaneous precipitation of mixed oxides, and Ewell and Insley (35b) co-precipitated alumina and silica gels. As noted earlier, the sol-gel technique was discovered in this laboratory. It has been virtually universally utilized for 30 years by the Penn State group and the many colleagues who participated in research in this group. It used the concept of total mixing in solution. Some tens of thousands of ceramic compositions in 2-10 component systems have been prepared since *1950* by mixing in solution and subsequently utilizing two main routes to solidification. First the *"sol-gel"* route. The general procedure involves mixing nitrates and metalorganics in alcohol water mixes, followed by gelation. Second, the *"evaporative decomposition"* route. The studies began with the work of Roy and Roy in the system $MgO-Al_2O_3-SiO_2-H_2O$ (36a), and have since been applied in almost every part of the periodic table. Relevant to the present compositions of mixed silicates, titanates, etc., was the early work by DeVries and Roy (36b), and Rase and Roy (36c) on the titanate systems using $TiCl_4$ and organic titanium precursors. Mumpton and Roy (37) used the techniques in the study of the $USiO_4-ThSiO_4-ThO_2-UO_2-ZrO_2-SiO_2$ system. Careful monitoring of total compositional changes in the sol-gel-desiccation process by Luth and Ingamells (38), showed that there was no change of composition, even in very complex systems. The sol-gel technique has been extensively applied in the nuclear industry by the ORNL in the making of oxide fuels. Furthermore, ORNL have recently developed a process by squirting the gel into $C_2H_3Cl_3$ (39) for obtaining a range of sizes of ceramic spherical beads. These ceramic beads can be handled much more cleanly than ceramic powders, and the process is more easily adapted to remote operation. The process (sol-gel route followed by ORNL bead making) has considerable merit for handling high concentration wastes which are to be made into ceramics. Indeed, processing into such beads may provide the lowest temperature route to making either ceramic or glass beads for incorporation in metal or concrete matrices.

Spray drying of nitrate effluent streams to a "calcine" has become a routine part of radwaste technology. This technique is a low temperature version of the technique of "evaporative decomposition" which was first used for these purposes in making fully anhydrous ceramic oxide powders by Majumdar and Roy (40); it was developed and commercialized to full scale (tons/day) in Austria by Ruthner (41). The difference is principally in the lower temperatures used, in spray drying, and hence in the latter's product which contains a variety of anions other than $O^{2-}$. Recently,

O'Holleran et al. (42) have carried Ruthner's work into the preparation of very homogeneous micron size ceramic powders of *complex* composition.

     a.  Ion-exchange Bed Adsorption.  A "mixing" strategy which has been a part of the radioactive waste technology from its earliest days has been the adsorption of ions onto organic or inorganic exchange resins.  The latter process of adsorbing on clays, zeolite, and gels in fact results in a very finely (atomic scale) mixed batch of fission products (from the liquid waste stream) and the desired hosts.  The silicate and alumino-silicate hosts were first noted in this regard by Hatch (8) and studied by Ames et al. (43a,b), decades ago.  The Sandia (17a,b) and Swedish (16) groups utilized $TiO_2$-rich gels and showed that the resulting precursor could be formed into homogeneous ceramics.  This method of mixing has a great advantage in its temperature of operation and following the route (see later) of mixed gels tailored to the final ceramic has great promise.

## Processing:  II, Consolidation and Densification

    A dozen different technologies have been developed over the last several decades for ceramic materials preparation.  We will discuss them in turn, with respect to their applicability (and application so far) to ceramic nuclear waste consolidation.

    *Liquid → Glass → Crystals* (glass-ceramic).  The most recently developed of all major ceramic technologies is the glass-ceramic route developed by Corning Glass Works in the mid-fifties under the trademark "Pyroceram." The process requires melting glass, casting or molding it into the final desired shape, then by *a series* of carefully controlled heat treatments, causing the formation of two unmixed glassy phases, which then promote nucleation of crystalline phases throughout the body (rather than on the surface only) and finally growth of the crystalline assemblages which, because of the large number of totally dispersed nuclei, yield a very fine microstructure.

    Research on this process for radwaste has been carried out at the Hahn-Meitner Institute in Berlin (44a,b,c) and right through to pilot plant in Germany.  Here, ceramic compositions with fission product loadings up to 30% and containing the following major phases have been prepared:  spinel, $Zn_2SiO_4$, $SrMoO_4$, Gd-titanates, $Ca_2Ln$-silicates, $RuO_2$, $CeO_2$.

    No "hot" tests have been made as yet, nor any scale-up beyond the several gm laboratory size ingots.  The pyroceram process is a relatively sophisticated one with respect to thermal control and cycling.  Some massive forms have been "cerammed"--e.g., a several ton block for a large telescope mirror--but the controls required are exquisite.  The product, of course, has the advantage of zero porosity and the crystallinity of the ceramic.  The compositional control, however, has to be compromised from the optimum to obtain the right crystallization behavior.  Only a few studies have been published regarding the distribution of the F.P. radio-nuclides among the phases (44c).

    *Liquid → Fine Grained Crystalline Block* (fusion-casting = controlled solidification).  This postwar development in metallurgy involves pouring the molten mixture into molds and by careful programming of the cooling causing the development of the appropriate microstructure to give the desired properties.  The typical ceramic melt will (except by chance) either form large crystals, or crack due to thermal stresses, etc., or form metastable assemblages (which could be desirable). Fusion-casting as a ceramic technology is used for making large shapes of very high temperature refractories. In principle, such controlled solidification would offer an alternative to the pyroceram process and one that could be more easily adapted to an in-can crystallization. In fact, in the early work on phosphate glass compositions, the materials obtained were often crystalline. The most relevant work perhaps was the WSEP program, which made a microcrystalline casting from a spray calciner melter at Hanford (45). Although not so labelled or planned, Ringwood's first ceramic compositions were also, in fact, "fusion-cast" (13). Solidification into the final crystalline phases is achieved directly in this process but the high temperatures often involved are undesirable. To date no purposive *process* research has been done on this major solidification technology.

*Solid T°C, Solid* (sintering). The vast majority of technical ceramics are densified by sintering. In radwaste ceramic preparation this process has often been implicitly assumed. Yet to date relatively little process work has been funded in the field. The Battelle PNL (46) have applied sintering after disc-pelletizing to McCarthy's tailored ceramics with considerable success. The supercalcine powder is formed into pellets and reacted at 1200°C for two hours. The product is a relatively porous bead, but relatively well crystallized. Normal sintering of most oxide ceramic compositions is a routine matter, and some optimization by utilizing various additives to give liquid phases, avoiding high temperatures which would volatilize some components, will yield a 90-95% dense body.

*Solid T, P(Uniaxial)*, (hot pressing). In the last two decades hot pressing has emerged as a routine technology in the production of certain specialty ceramics. The method consists of sintering at modest temperatures (up to 1200°C) at modest pressures (<100 MPa) for short periods (∿1 hour). Automated hot pressing is used in the production of ceramic capacitors--where relatively small objects are made in large amounts. It is also used in producing specialty (transparent), theoretically-dense 12" diameter x 2" thick blocks of $Al_2O_3$, spinel, etc.

Hot-pressing was the first technique used by McCarthy et al. (11,29), in laboratory consolidation of their ceramic wastes from 1973 onwards. As a process it has certain advantages and disadvantages. The latter include: It is more of a batch process than most; it handles relatively small units; there is little experience with remote operation to date. The former includes: It is carried out in a closed system and will suppress any volatilization; it yields a reproducible dense product at lower temperatures than any L→S or sintering process. It is an ideal process for the *laboratory* evaluation of different *compositions*, since one can produce a uniform dense body the first time, every time (if compatible with die material).

*Solid T, P(Hydrostatic), Solid* (hot isostatic pressing). Hot isostatic pressing of large shapes is a rather new technology. Although obviously related to hot pressing it has certain major advantages for radwaste applications:

    (1) It can handle *much* larger units.
    (2) It is often operated "remote," because of the danger from the stored energy in the compressed gas.

(3) It could provide densification, encapsulant, and container formation all in one step.

Isostatic pressing requires a "can" or "bag" of deformable material. This is typically a metal for ceramic shapes, although bags of viscous glass have very recently been developed. The deformation limits of the metal often require that there be two steps in the isostatic pressing, since powder packing densities are relatively low. The most important radwaste research with this process has been carried out at ASEA at Robertsfors in Sweden. Here in a two-step process, a ceramic waste assemblage was encapsulated in a ceramic (mullite) and encased *and sealed* in a ceramic container (polycrystalline sapphire) with a full scale size of 3m x 1m diameter. A batch process at this scale offers no major disadvantage where throughput per unit-time is of secondary importance. Hot isostatic pressing has many obvious attributes for radwaste processing and will likely appear more prominently in future options.

## Coated Ceramic Radiophases

A further significant development in ceramics has been the demonstration of the ceramic pellet doubly coated with layers of pyrolytic C, followed by $Al_2O_3$. Rusin et al. (47a), Bonner (47b), and McCarthy (47c) have described the preparations of these disc-pelletized tailored ceramics of compositions corresponding to SPC-2, and their subsequent CVD coating. In the comparative studies of fully loaded waste forms (glass, uncoated ceramics, spent fuel, etc.) in various host rocks, these doubly-coated ceramics have come very close to reaching the "zero-release solids." In hydrothermal runs, even at 400°C in contact with any silicate rock, where all other forms are 50-100% altered in a day, the coated ceramics show no attack after two months (47c).

## Hydroxylated Radiophases

Roy (48) has recently carried further the identical argument as that advanced by McCarthy and Roy and supported later by Ringwood, that crystalline phases (both radio- and encapsulant) are closer to equilibrium and hence will survive with less alteration in this repository environment. He has shown that under the newly proposed repository conditions (100 bars 200°C) anhydrous ceramic radiophases must be less stable than hydroxylated ones. Equilibrium phase diagrams of all the major oxide binary systems and many ternary and quaternary systems with water are known. They show that except for the tetravalent oxides ($SiO_2$, $GeO_2$, $TiO_2$, $ZrO_2$, $UO_2$ and their compounds such as $ZrSiO_4$ (= $CePO_4$), all oxide assemblages will hydrate *to the less soluble* hydroxylated phases. Typical radiophases are dioctahedral Mica (Cs), wairakite (Sr, Hydroxy-apatite (Sr,Ln,An), Sodalite (Sr, Rb), nepheline hydrate I, montmorillonite-vermiculite (Cs,Sr,Ln), zeolite ($Lr^{3+}$,$An^{3+}$). Although few such experiments have been completed, an in-can reaction at approximately 100°C is envisaged built on the model of the Oak Ridge "Fuetap" process (see below).

## Spent Fuel

Spent fuel was never considered a waste form till April 1977, when President Carter proposed that the U.S. use the once-through fuel cycle. Thus, the information on this ceramic radiophase waste form itself is very small. This topic will also be treated in detail in a later report in this series.

a. Compositions of Waste. There have been no "hot" optical microscope or x-ray diffraction data published on real spent fuel. Hence, we do not as of this date know for sure what phases are present in it. However, it is virtually certain on crystal chemical grounds that a single phase *predominates*. This is a fluorite structure phase with the composition roughly as follows: $[U^{4+} \cdot Pu^{4+} \cdot all\text{-}An^{3+\&4+} \cdot all\text{-}Ln^{3+\&4+} \cdot U^{6+} \cdot Pu^{6+} \cdot Sr^{2+} \; O_{2-x} \; \Box_x]$. The principal ions which will not be accommodated in this structure will be $Cs^+$, $Mo^{6+}$, $Ru^{4+}$, $Tc^{4+}$, $I^-$, Xe, and Kr; the latter two will be gases.

b. Waste Forms--Radiophases. For spent fuel rods which *may* become the chief waste form of the future, the only seriously-considered strategy is the encapsulation strategy. This will mean leaving the fluorite radio-phase totally untouched (see below).

However, at the TMI plant in Harrisburg, it is highly probable that such a fluorite oxide *waste* (*not* spent fuel rods) diluted with $ZrO_2$, exists in amounts in the 1-10 megacurie range. No research or conceptualization of the solidification options for this unique waste has been started. This will present an intriguing challenge to the materials scientist, and optimum design of radiophases and encapsulant phases for this situation are highly unlikely to be similar to those used, let us say, for defense wastes or hot commercial liquid reprocessed wastes.

## Summary on Status of Ceramic Radiophase Research

From the decade or so of work at Penn State and the HMI-Berlin work (loc. cit.), the viability of commercial fission product incorporation at high concentration in a fully crystalline waste form, has been established. *Dilute* solutions of fission products are of course possible in these, but also in a wider range of phase assemblages. Especially for the low concentration radwaste case, a virtually infinite set is available and optimization will require some rational strategy yet to be enunciated.

An area in which insufficient work seems to have been done is in the "low"-temperature ceramic-radiophases. Consolidation processes which could operate at low temperatures--even room temperature--offer *enormous* process reliability advantages in a remote operation compared to operation at 1200°C. However, there has been very little research on the crystalline radiophases which can be formed at such low temperatures. The opportunity for instance, of Sr and Cs incorporation in a mica using as a precursor template a vermiculite or dioctahedral montmorillonoid has only been explored in a very preliminary way. Likewise, preliminary research on cancrinite-sodalite plus layer-silicate assemblages seems to have been suspended, although their processing advantages may outweigh product inferiorities.

Ceramic *processing* research on radwaste is just beginning. Sintering, fusion casting, hot pressing, hot isostatic pressing and hydrothermal autoclaving all appear to be well within the range of technical feasibility. Detailed work on specific compositions at this stage would appear to be wasted, since compositions could well change drastically. What is needed is the demonstration of reliability and simplicity of densification on a class of compositions.

# Chapter 6.  ENCAPSULATED WASTE FORMS

## *Synopsis*

*The strategy adopted in the ceramic radiophase approach described in the previous chapter is to make a multiphase ceramic, where each of the phases containing radionuclides is highly insoluble or unreactive at modest $pH_2O$ and $t$.  The encapsulated waste form is based on a very different idea, that one can physically surround radiophase particles whether highly leach resistant or not by very inert ceramic (or other) phases.  Obviously this strategy will work best when the concentration of radiophases is low. Thus the encapsulated strategy rests on the high <u>impermeability</u> of the assemblage and the chemical resistance of the encapsulating phase.*

*The figure below will help define the difference in concept.*

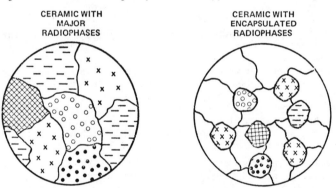

*Three categories of encapsulating materials are discussed: ceramics, metals, and cement.  Of these the former two represent examples of high technology, relatively complex processing suitable for commercial waste. The cement matrix on the other hand is an enormously significant technology.  It has repeatedly been recommended by Advisory groups (such as the U.S. National Academy of Sciences committee on Radioactive Waste Management, for the immobilization of wastes with low concentration of radiophases, such as defense wastes.  It represents the only waste form actually used in waste disposal (at Oak Ridge).  Its importance for national and international waste disposal can hardly be overstated, even though it has and continues to be neglected.*

# ENCAPSULATED WASTE FORMS

The balance between the relative reliance on solid phase immobilization and isolation has changed towards the former. We believe that it is certain that a second change in balance will occur at the next level down in aggregation of the system. The last few years have seen a great increase in concern for the stability and insolubility of the radiophases as described above. It is our judgment on the basis of purely scientific considerations, that the balance will now shift towards the encapsulant phases and microstructure of the encapsulant + radiophases combined. In the light of this, the following sections take on an added importance. This is especially true for defense wastes where location and cost considerations have made cement encapsulated wastes an obvious favorite.

## Ceramic-Matrix Materials:  The Concept

The concept of the ceramic matrix is to provide a high resistance oxide material as a physical barrier between any infiltrating solution and the radiophases. It is illustrated in the figure above. Crystalline mineral-modelled phases have had their stability or at least, persistence, as encapsulants in various near-surface ambients tested in nature for millions of years. In dense crystalline rocks we frequently find "weatherable" phases, encapsulated or surrounded--and hence protected by the less weatherable phases. It is difficult to conceive of ordinary geological or seabed environments where a dense ceramic would not be considerably more resistant *as a whole* than any radiophase glass.

Especially in the case of low concentration wastes where the existing diluent material is already a resistate ceramic phase, the advantages of this waste form in savings in processing complexity are immediately obvious. In high concentration wastes, one can make an encapsulated waste with an inert outer wall of pure ceramic encapsulant, thus providing in effect a secondary ceramic "container" around the encapsulant phases.

## The Compositions Studied

a. High F.P. Concentrations. It does not seem to be widely appreciated that a large number of ceramic matrix waste forms have, in fact, been proposed and made up, and evaluated. We have alluded to some of these under the ceramic radiophase section, since in many cases no sharp division is possible. The earliest work among those developed through a processing stage was that of Arrance (9) in 1963. The concept was of a pre-formed porous ceramic shaped into bricks or balls, which were used to adsorb the waste-stream ions (and solids?). The ceramic was fired to form "mineral-like phases" and the whole shape could be overglazed to zero porosity. The patent reported very few details. In 1973-1976, McCarthy et al. (48a) explicitly described the ceramic-matrix encapsulation. They applied it to mixtures of calcine-ceramic powders with quartz as the inert matrix material, which were hot-pressed together. These ceramic-matrix materials had leach resistances comparable to those of Battelle glasses and some ceramic hot pressed radiophases. McCarthy (48b) studied some ten encapsulates including quartz, mullite, and $Al_2O_3$. In 1976, he reviewed the work on "matrix isolation," showing that calcine hot pressed at 1100°C with quartz (and ∿4% glass) gave a product as good as the current radio-phase glass (11d). Some years later, Larker (49) hot isostatically pressed McCarthy's tailored-ceramic powder in a mullite matrix, making a very satisfactory product.

Two groups in the middle- and late-seventies took a different process and a different composition approach to ceramic matrix composites. The differences are clearly highlighted in Table 12. Schwoebel and Johnstone (17a,b), and Dosch (17c), in the Sandia group, have pioneered in the use of $TiO_2$ gels as ion exchange resins and adsorbers for the typical commercial waste. They have in fact carried the process through with real wastes. During the course of the years, they had broadened the gel composition to include niobates and zirconates made by the hydrolysis of organics, as in the ceramic sol-gel process, and included crystalline zeolites as adsorbers for $Cs^+$. This is a most attractive process for development, since it involves the lowest temperatures for precursor formation.

The Swedish group [Forberg et al. (17)] who first categorically used the term microencapsulation modified the adsorber bed to adsorb the typical HLLW stream rather efficiently. However, it is not clear what their maximum fission product loading was, since $TiO_2$ is the predominant phase with only perovskite as an observed second phase in the x-ray pattern of their fired product.

The Sandia group has used principally hot-pressing for consolidation, whereas the Swedish team has used hot isostatic pressing after dehydrating at 700°C. The latter provides the lowest temperature route to ceramic matrix encapsulation.

b. Low F.P. Concentration. Most recently, this same concept has been applied a fortiori to defense waste sludges. Inspection of the compositions of these sludges showed that they already have compositions which would have to yield upon firing, a mixture of two extremely resistant structures, the $R_2O_3$-corundum and the $A_2BO_4$-spinels as the major encapsulant phases if they were hot pressed with only minor additives (50). The principle here is to use the sludge composition itself as the encapsulant to the maximum extent possible. Most recently, Morgan et al. at Rockwell (51a), and Roy et al. (34) have made other minor additions and hot pressed some simulated Savannah River calcines as well as sludges which also contain magneto-plumbite as a radiophase. Al and Fe, dominant ions in the sludges, substituted interchangeably in both structures of the dominant encapsulant phases, corundum and spinel.

c. Processing. Relying on ceramic encapsulation is the fundamental concept behind the shortening of the total process line for defense wastes. It eliminates any $Al_2O_3$-removal step, possibly an alkali removal, and

## TABLE 12

### Ceramic Matrix Waste Forms

| Encapsulants Phases(s) | Radiophases where known | % F.P. in total body | Authors or Reference | Process |
|---|---|---|---|---|
| Ceramic "sponge" principally meta-kaolin, mullite, Etc. | unknown | not defined | Arrance, 1963 | Adsorb on pre-forms + fire |
| Quartz (and 9 others) | fluorite, perovskite, apatite, various silicates | ∿20% | McCarthy, 1973, 1976 | Mix powders hot press |
| Mullite + $Al_2O_3$ | SPC-2 set* | 20% | Larker, ASEA, 1979 | Mix powders hot isostatic press |
| $TiO_2$ | Perovskite + ? | Few % | Forberg et al., 1979 | Adsorb on gel bed, hot isostatic press |
| $TiO_2$ | a. pollucite $Gd_2Ti_2O_7$ b. fluorite, amorphous, etc. | ∿20% | Dosch, 1979 | Absorb on gel bed, hot press |
| $Al_2O_3$, spinel, etc. | unknown + mag-netoplumbite | <1% | Morgan et al., 1980 | Slurry + additive; hot press, etc. |

*See Table 10 and reference (29).

possibly the dissolution and spray calcining steps. Many of the radiophases have not yet been identified, although one of them even if dilute in radionuclides, as noted above, has been found by the Rockwell group to be magnetoplumbite.

As noted, processing of the $TiO_2$-matrix phases for high concentration wastes has been carried through laboratory and even to hot experimental stages. The quartz matrix materials have been made in 2" diameter pellets, also by hot pressing.

At the other end of the spectrum is the ASEA work where hot isostatic compression has been carried out both on laboratory-scale samples for $TiO_2$-matrix and mullite-matrix, and at full scale to make the largest *simulated* rad-waste specimen in the world.

d. Properties. From the mineralogical record, the long-term stability of especially an $Al_2O_3$ or mullite-matrix body in a silicate host rock will be found to be outstanding. Especially since the original concentration of radio ions is <1%, the leaching characteristics will very likely fall within ±1 order of magnitude compared to glass or concrete. Experimental data are now needed on these properties.


## Metal-Matrix Materials

### Importance of Thermal Effects in Waste Forms

The original concept of the metal matrix was to provide a macroscopic (i.e., a 1cm scale) encapsulation of the glass beads which constituted the radiophase. The matrix here provided both a chemical barrier as well as a high thermal conductivity sheath for lowering the temperature of the composite in certain configurations. Such a concept may, of course, be generalized to much finer sizes of radio- and encapsulant-phases as has recently been done.

### The Metal Compositions Involved

a. High Concentration Wastes Including Spent Fuel. This concept has been translated into a reality over many years of research at the Belgium Nuclear Center at Mol. Van Geel and Detilleux (52a,b) have described these materials in detail, and the pilot plant has been working for some years. Here the metal matrix is a continuous phase of lead on the order of 1 meter in size, into which cm. size radiophase phosphate glass beads have been introduced. Other laboratories (53) have duplicated this effort, almost all relying on low melting alloys--chiefly of lead, possibly alloyed with Sb or Sn because the glass beads would otherwise flow together.

At Oak Ridge recently (54), the use of Fe-Ni alloys has been suggested. Laboratory experiments have led to a Fe-Ni metallic matrix material, but with a very different microstructure of micron size aggregates of a calcined ceramic radiophases. Optimization of the ceramic along McCarthy's tailored ceramic lines is proceeding.

By far the most significant use of metal encapsulants is in the proposed use of lead to encapsulate spent fuel rods. This use is very likely to be used in any program which actually disposes of fuel rods.

b. Low Concentration Wastes. The only metal proposed for encapsulation of defense wastes has been copper, by Macmillan et al. (19).

## Processing

The Belgium lead-matrix phosphate glass beads in lead process has been engineered through the hot pilot plant stage. The process has one major advantage over bulk glass processes, *in that the product radiophase is continuously inspectable*. Indeed, it may be sampled from time to time. The lead infiltration process is simple and involves a low temperature.

The new Oak Ridge cermet process is much more complex. It involves reduction of the Fe and Ni in situ around the ceramic calcine by reduction in hydrogen at a high temperature, followed by hot pressing or sintering (at 1200°C) (54). W. Neumann (53) has examined some new processes for mixing metals and calcines, including "thixo" casting. These are innovative appraoches which *may* produce effective new processing technology.

The Penn State electrochemical process (19) is unique in that the entire waste form processing consisting of incorporating defense sludges into a copper matrix has been carried out *at room temperature*. It has the additional advantage of the greater availability and lower cost of copper, as compared to lead. The amounts of metal required may prove to be prohibitive for many metal matrix applications, especially for high volume wastes. Moreover, this work is far from optimized even in the laboratory.

One of the earliest processes--the *low temperature solidification (LOTES)* process--is a hybrid of a ceramic-matrix and a metal-matrix process [Van Geel and Eschrisch (52a,b)]. Here the calcine is incorporated *onto* an AlPO₄ granule. Such a material would be very similar to McCarthy's calcine + quartz matrix. Much of the LOTES work, however, went further and incorporated 35% of the ceramic-matrix granules into metal. There is little question but that such an optimized (super)calcine + ceramic matrix in metal matrix would represent an extremely insoluble, thermally and chemically stable, waste form. What is most significant is that these forms have been completely engineered for some years.

The KBS-Study (55) presented to the Swedish parliament as a *PLAN* for waste storage, proposed that spent fuel rod assemblies--including their zircaloy cladding--be encapsulated in lead. The entire set of rods with the massive lead encapsulation, would be contained in a copper canister. In effect, here (just as in the extensive, already solidified ceramic Hanford "needles" or capsules), we have not so much an encapsulant and canister, but a double canister (see later section).

## Properties

It is not a straightforward procedure to measure the relevant properties of an encapsulated-in-matrix material when the radiophase is large in size. With no glass beads exposed, the Eurochemic Pb-matrix waste form naturally exhibits the leaching properties of lead, i.e., zero release of radionuclides even at the maximum design temperature in the presence of water, but finite dissolution of lead. To characterize the total waste form it is obviously mandatory to have solubility and reactivity data on the radiophase also. The impact resistance of the full scale containers has been shown to be excellent. Neumann (53) has summarized some of these properties of calcine encapsulated in metals. No *significant* property measurements have been designed or carried out on the other examples as yet.

## Glass-Matrix Materials

The Concept

This is exactly analogous to the ceramic matrix described above, except that the matrix is noncrystalline. Figure VI and Table 13 give a quick general picture of what is involved. However, the glass matrix form differs at least in degree from the metal and ceramic encapsulants in one respect. The metal is for all intents and purposes, totally inert with respect to the typical radiophases incorporated--it does not itself become a radiophase. However, in a multiphase ceramic encapsulant, while some phases such as well crystallized corundum and quartz, are in effect quite inert under the processing used, others will react with the radionuclides, and one gets an intimate micron-scale intergrowth of radiophases and inert phases. In the glass case--and to some extent the cement case, a part of the glass is likely to be reactive, but much of it is inert. It should, however, be noted that in the limit even lead reacts with many ceramic phases, so that not all metal matrices are inert.

Compositions Studied and Processes Used

Table 13 provides a birdseye view of the state of this subject. It will be noted that in 1972 McCarthy and Lovette (48a) reported on the mixing of borosilicate glass with 20-30% waste calcine powders, and the hot pressing at temperatures of 500-900°C and pressures of 1000-4000 psi into cylindrical shapes. In 1975, W. Ross (56) reported a rather similar approach, using roughly the same compositions but using *sintering* instead of hot pressing. At the same meeting, McCarthy and Davidson (11b) reported again on making matrix-encapsulated samples using ten different encapsulants, including a mixture of a ceramic (quartz) and a lead borosilicate glass encapsulant. Their best samples were obtained with a Ludox encapsulant, yielding silica glass + cristobalite in the compact. In 1979, Macedo et al. (57a,b) modified the Ross approach by increasing the internal surface area of the borosilicate. The concept they used was that a leached borosilicate glass frit in addition to the encapsulant action was a sponge to adsorb and react with some ions. No specific data on waste-loadings and resultant phases or properties were provided. A change from Ross' work was the consolidation by sintering after the powder was packed into a high silica glass tube which served as a "container," thus creating an extra macro encapsulating layer.

This process appears to be the successor to a slightly different concept described by Simmons et al. (58) which was in turn a successor to Arrance's ceramic sponge (9). Instead of Arrance's metakaolin--noncrystalline aluminosilicate preforms--Simmons et al. used leached borosilicate--Vycor-like--preforms for the same "soak and dry" process to

TABLE 13

Metal and Glass Matrix-Dependent Forms

| | Encapsulant | Radiophases(s) | Authors* | Description | |
|---|---|---|---|---|---|
| **METAL** | Lead | Glass Ceramics Calcines | van Geel et al. | Phosphate (silicate) glass beads in lead matrix | Major highly developed technology |
| | | | Neumann | New processes to achieve same end | Little work done |
| | Fe-Ni | Calcine Ceramic | Aaron, Quimby et al. | Micron size radio-phases; metal formed by in situ reduction | Very preliminary work |
| | Copper | Waste Sludge | Macmillan et al. | Electro co-deposition at room temperature | Very preliminary work |
| **GLASS** | Borosilicate glass, Ludox, etc. | Calcine-ceramic | McCarthy and Lovette | Hot pressing | Very preliminary work |
| | Lead boro-silicate glass + quartz | Calcine-ceramic | McCarthy and Davidson | | |
| | Borosilicate glass | Calcine (?) | Ross | Sintering to consolidate | |
| | Porous (leached?) borosilicate | Low conidation | Macedo et al. Simmons et al. | Sintering with outer high silica glass sheath | Modest amount of data available |

*See text for references.

adsorb and "fix" certain specific *ions in solution*. They also obtained a "glazing effect" of an outer less contaminated skin, as Arrance had done by another route. This was done by washing off the outer layers, and collapsing the structure at 700-900°C. They obtained a graded composition with only ~5% radio-ions in the outer layer and ~20% in the inner areas. On sintering, this gives a high-silica skin. The data presented so far are very preliminary. For instance, they deal only with the incorporation of a single ion (Cs) at a time not any standard combination of FD ions.

## Properties

Few experimental data on chemical properties have been reported on the representative radiophase + *encapsulant phase(s)* exposure to leaching conditions by any of these authors. Macedo et al. (57) have presented data on the encapsulating high-$SiO_2$ glass sheaths which, as may be expected, show very low ($SiO_2$) solubilities of ~$1\times10^{-6}$ g/cm$^2$ day, in water at 100°C, but say nothing whatever about the solubility or leachability of their radiophases.

# Cement-Matrix Materials

## Importance

The importance of this particular waste form in the national, indeed the worldwide radioactive waste disposal strategy is enormous. Especially for dilute wastes such as low and medium level wastes worldwide and U.S. defense wastes, the role of cement encapsulated radiophases is certain to increase. Statements such as the following are typical after analysis of the options by experienced deliberative bodies:

> "It has been claimed that incorporation of commercial high-level and defense wastes into a cement matrix has resulted in products essentially equal to the typical glasses in leachability and thermal conductivity. The potential advantages of incorporation into cement are the simplicity of the process, the lower temperatures and the possibility of on-site disposal."*

The continuing neglect of this option by decision makers everywhere is evidence of technological momentum generated by a combination of ignorance of or a lack of understanding of the data, fear of charge from an apparently working solution, vested interests in maximum cost solutions etc.

## The Concept

The cement-matrix materials constitute a subset of the ceramic-matrix materials. However, they differ in one major respect and in some details.

---

*Outlook for Science and Technology: The next five years, USA National Research Council, April 1981 *(Prepared by the Committee on Radioactive Waste Management, N.A.S., Freeman, NY)*.

The major difference is that there is already substantial backup of research papers, demonstrated solidification and disposal systems, and hence this group merits considerably greater attention. It is also the only system which does not require a container at all, and which potentially eliminates transportation altogether. Hence, it is in many ways a unique SPI subsystem with major potential advantages. The cement matrix waste form also differs from the ceramic matrix form in being partly crystalline and partly amorphous. It is also the only hydrated matrix-- which may, in fact, be closer to equilibrium than the equivalent dry material in a water-saturated rock.

## Compositions Studied

Several groups of workers have studied cement-matrix, intermediate to high level waste encapsulation in cement. Oak Ridge National Laboratory has pioneered the use of a concrete grout for injection directly into a hydrofractured shale rock. They have used ordinary portland cement mixed with a variety of additives, as shown in Table 14. The total process has been described recently by H. Weeren (59). In other more recent work, clay and other additives are used to fix $Cs^{1+}$ and other ions and the cement (+ additives) slurry is autoclaved in its own final container [Moore et al. (60,65a,b)].

At the Brookhaven National Laboratories, considerable work was done for years by Colombo and co-workers (61), to utilize polymer-concretes for enhanced impermeability and strength. Stone (62) at the Savannah River Laboratories, explored the making of cement waste composites using traditional technologies, and carried out solidification of actual waste sludges. At Penn State, on the other hand, unconventional approaches to cement matrix solids have been taken by D.M. Roy (63,64a,b) and her co-workers. First, ultradense cement composites were made by hot pressing waste-cement mixtures, or encapsulating simulated waste calcines as a core inside an outer cement shield. Second, a systematic study of the reactions and stability of the individual most significant radiophases in high alumina and regular portland cements has been undertaken [Roy et al. (64b)].

However, the most radical departure from the use of calcium silicate based cements was achieved in the Eurochemic plant in Mol, where hot cladding hulls with considerable fission product and transuranics contamination were incorporated into an alumino-silicate cement in a 25cm plastic column. Penn State work on phosphate and other cements is a parallel development which seeks to expand the range of chemistries and setting times available in state-of-the-art low temperature encapsulation technology.

## Processing

a. Grouting. A grout incorporating waste liquid or dried sludge is formed by blending with cement and other additives and this slurry is injected into beds in a shale formation. The shale is first hydrofractured with water, and as the crack propagates, is followed by the grout. Injection continues until a batch is depleted; and the emplaced grout sets within a few hours. Subsequent injections are carried out later in sheets parallel to and above the first or preceding one (15).

## TABLE 14

### Cement-Matrix Waste Forms

| Compositions Including Additives | % Fission Products + TRU | Processing | Cements | Reference |
|---|---|---|---|---|
| (a) Portland cement + Conasauga shale, clay, fly ash | <<1% | Grouted in place in shale | Monitoring shows negligible RW release | Weeren (59) (1977), |
| (b) Portland, high alumina cements, clay, fly ash, $NaNO_3$ | <1% (<15% dried, simulated sludge) | Autoclaved at saturated vapor pressure | Low Cs leachability; radiolytic gas recombination | Moore (65) (1979), Oak Ridge |
| Portland, high alumina cements (+ zeolite) polymers (+ vermiculite) | <1% (<50% dried, simulated sludge (<15% zeolite) | Normal ambient solidification; monomer soak; cure | Impermeability of polymer concrete | Colombo and Nielson (61) (1977), Brookhaven |
| Portland, high alumina cements include actual dried sludges + zeolites, retarder | <40% dried sludge <2.5% Cs-zeolite | Normal ambient solidification | High alumina cements superior; low α-emitter, Sr leachability. Real similar to simulated; high strength | Stone (62) (1977), Savannah River Lab. |
| Portland, high alumina, and modified cements (a) + calcines, supercalcines (b) + individual radiophases | 20% HLW calcine or supercalcine 10-30% | Hot pressed (150°C) Regular set | Also encased in cement container. Hydrothermal leaching studies, low RW release | Roy and Gouda (66) (1978) Roy et al. (1979a,b), PSU (64a,b) |
| $Al_2O_3$ + $SiO_2$ gels + polymer "concrete" | Cladding hulls | Room temperature set | Compaction of hulls | Eurochemic, Belgium (67) |
| Cement concrete | Cladding hulls | Regular set | | G.f.K. (68) Karlsruhe (CEC, 1977) |
| High alumina cement concrete | Calcine pellets <60 wt % calcine | Regular set | Strong to 250-275°C Leach resistant | Berreth (69) (1976), INEL |

b. Room Temperature Conventional. Cement, additives, water, and dry or liquid waste solid wastes may be contained in pre-pelletized forms, are blended in a planetary, rotary, or pan mixer or other suitable type of apparatus, and poured into molds. The latter may be self-molds (concrete, polymer impregnated concrete, or reinforced concrete, without waste), or metal canisters. Optionally, dry ingredients may be pre-blended, and the wet and dry ingredients added simultaneously to a can, which is capped and rotated and possibly vibrated to blend individual batch lots. Curing includes normal ambient or, alternatively, slightly elevated temperatures, <100°C.

c. Autoclaving. This process is akin to the commercial process for making autoclaved cement block and related building materials. It is a simple technology, widely used in industry and hence suitable for remote operation. Cement and additives are blended with dry waste solids and water, and cast in heavy walled cans. A ventable lid is fixed in place, the can is sealed, and autoclaving takes place at temperatures up to 250°C at saturated steam pressure. After about one day curing, the can is vented, and temperature maintained at 250°C for ca. 48 hours to dry the concrete (65a). (This prevents the problem of radiolysis of *free* water.)

d. Hot-Pressing. Cement and other ingredients are blended, water (if required) is added, and the mixing continued. In one option, the solid ingredients contain sufficient water without further addition. The batch is formed in a lubricated die and is hot pressed at temperatures from 150-250°C at pressures of 10,000-50,000 psi, for typically 0.5 hours, then cooled (65). No drying is required, and the hardened monolith is extruded from the die into a canister.

## Properties of Cement-Matrix Materials

Considerable ignorance exists regarding the properties of modern cement composites. Grave misunderstandings of the role of water, the thermal stability, etc., are evident in the literature. Hence the properties should be compared with other waste forms.

The field monitoring at Oak Ridge from wells and streams of actual radioactive grout sheets (1,70a,b) has shown very little loss of any radionuclides. This is, in a way, an ultimate composite test of the satisfactory nature of the product.

Moore et al. (60), Stone et al. (62), and D.M. Roy (64) all report that leach tests on cement matrix forms containing 20-30% waste show generally comparable values to those for U.S. reference glass within an order of magnitude. Table 15 below shows a recent summary by D.M. Roy and co-workers comparing the "leach" rates of glass, and cement matrix formes incorporating Savannah River wastes. Although not strictly comparable because the data are from different reports, *it shows clearly that a concrete-encapsulated waste form can be made which is essentially as "leach resistant" as glass.* What is even more important is that in the planned repositories the cement phases will be much closer to equilibrium (hence less dissolvable) with the likely backfill silicate host rocks.

## TABLE 15

### Comparative "Best" Reported Leach Rates for
### Cement Matrix SRP Waste Forms ($g/m^2d$)
### Contrasted to a Reference Glass

|  | "Ref" Glass[a] | Cement Matrix Composites | | | |
|---|---|---|---|---|---|
|  |  | SRL[b] | ORNL[c] Fuetap | 60° adj. (Aluminate) | Warm pressed |
| Waste loading | 30% | 40% | 15% | 20% | 40% |
| Cesium | 2.7 | 10 | 20 | <.1 | <.1 |
| Strontium | 0.2 | 0.1 | 0.45 | <.7 | 0.6 |
| Leach test | MCC-1 | Paige | MCC-1 | Paige | Paige |
| Nominal temp. | 90°C | 25°C | 90°C | 25°C | 25°C |
| Test duration | 28d | 42d | 28d | 25d | 7d |

[a] A. Harker, personal communication
[b] Stone, J., DP-1448, I.E. DuPont de Nemours, Aiken, SC (1977)
[c] Moore, Ref. 65
Other cement data from D.M. Roy, The Pennsylvania State University

 

Thermal conductivities are also not very different (cf. 0.4 to 0.6 w/m°K for FUETAP, 0.7 for glass (Moore, 65a), and about 0.8-1.0 for hot pressed concrete). Strengths for normal concretes are lower than those of glasses, but frangibility and tendency to powder upon impact may be no worse. Thermal stability between 250° and several hundred °C shows that it is comparable with other waste forms.

One of the major questions which have been raised about the suitability of cement as a matrix, was the possibility of radiolysis of the $OH^-$ groups in the calcium silicate hydrate phase of set portland cements. J. Crawford, writing in the NAS study on waste solidification (1), summarizing the state of the art, discounted the likelihood of such radiogenic gas formation being a problem, on the grounds of the rapid recombination of free radicals in such high surface area solids. *Earlier work in which the concrete had not been dried before sealing, has apparently produced misleading results on pressure buildup.* The most recent work at Oak Ridge (Moore, 65b) present data to show that the problem of radiolysis of properly dried cement encapsulated wastes is non-existent or very minor.

# REFERENCES FOR CHAPTER 6

1. National Research Council, Committee on Radioactive Waste Management, Panel on Waste Solidification (1978), Solidification of High Level Radioactive Wastes, National Academy of Sciences, Washington, DC.

2. R. Roy (1979), Science Underlying Radioactive Waste Management: Status and Needs, Scientific Basis for Nuclear Waste Management, Vol. 1, Ed. G.J. McCarthy, Plenum Press, NY, p. 1.

3. R.Roy (1978), Stabilities of Alternate Waste Forms, Proc. Conf. High-Level Radioactive Solid Waste Forms, Ed. L.A. Casey, U.S. Nuclear Regulatory Commission, Denver, CO, p. 757.

4. G. Jenks (1977), National Waste Terminal Storage Program Conference on Waste-Rock Interactions, Report No. Y/OWI/SUB-77/14268, The Pennsylvania State University, University Park, PA (See especially papers by W.B. White, D.M. Roy, G. McCarthy, and G.W. Brindley).

5. Committee on Radioactive Waste Management, Panel of Engineered Storage, National Research Council, National Academy of Sciences, Washington, DC (1975), Interim Storage of Solidified High Level Waste.

6. G.J. McCarthy, W.B. White, R. Roy, B.E. Scheetz, S. Komarneni, D.K. Smith, and D.M. Roy, Interactions Between Nuclear Waste and Surrounding Rock, Nature 273, 216 (1978).

7a. J.E. Mendel (1978), The Storage and Disposal of Radioactive Waste as Glass in Canisters, PNL-2764/UC-70, Battelle, Pacific Northwest Laboratories, Richland, WA.

7b. J.E. Mendel, W.A. Ross, R.P. Turcotte, and J.L. McElroy (1978), Physical Properties of Waste Glass, Proc. Conf. High-Level Radioactive Solid Waste Forms, Ed. L.A. Casey, U.S. Nuclear Regulatory Commission, Denver, CO, p. 37.

7c. W.A. Ross, D.J. Bradley, L.R. Bunnell, W.J. Gray, Y.B. Katayama, G.B. Mellinger, J.E. Mendel, F.P. Roberts, R.P. Turcotte, J.W. Wald, W.E. Weber, and J.H. Westsik, Jr. (1978), Annual Report on the Characterization of High-Level Waste Glasses, PNL-2625/UC-70, Battelle, Pacific Northwest Laboratories, Richland, WA.

8. L.P. Hatch, Ultimate Disposal of Radioactive Wastes, Am. Scientist, 410 (July 1953).

9. F.C. Arrance (June 11, 1963), Method for Disposing of Radioactive Waste and Resultant Product, U.S. Patent 3,093,593.

10. R.E. Isaacson and L.E. Brownell (1972), Ultimate Storage of Radioactive Wastes in Terrestrial Environments, Management of Radioactive Wastes from Fuel Reprocessing, OECD Proceedings, Paris, p. 953.

11a. G.J. McCarthy, Quartz-Matrix Isolation of Radioactive Wastes, J. Mat. Sci. 8, 1358 (1973).

11b. G.J. McCarthy and M.T. Davidson, Ceramic Nuclear Waste Forms: I. Crystal Chemistry and Phase Formation, Amer. Ceram. Soc. Bull. 54, 728 (1975).

11c. G J. McCarthy, High-Level Waste Ceramics, Trans. Amer. Nucl. Soc. 23, 168 (1976).

11d. G.J. McCarthy and M.T. Davidson, Ceramic Nuclear Waste Forms: II. A Ceramic-Waste Composite Prepared by Hot-Pressing, Bull. Amer. Ceram. Soc. 55, 190 (1976).

11e. G.J. McCarthy, High Level Waste Ceramics: Materials Considerations, Process Simulation and Product Characterization, Nucl. Tech. 32, 92 (1977).

12.  R. Roy, Ceramic Science of Nuclear Waste Fixation (Abstract), Bull. Amer. Ceram. Soc. 54, 459 (1975); and Rational Molecular Engineering of Ceramic Materials, J. Amer. Ceram. Soc. 60, 350 (1977).

13a. A.E. Ringwood (1978), Safe Disposal of High-Level Nuclear Reactor Wastes: A New Strategy, A.N.U. Press, Canberra, Australia.

13b. A.E. Ringwood, S.E. Kesson, N.G. Ware, W.O. Hibberson, and A. Major, The SYNROC Process: A Geochemical Approach to Nuclear Waste Immobilization, Geochem. J. 13, 141 (1979).

14.  V.I. Spitsyn and V.D. Balukova (1979), The Scientific Basis For, and Experience With, Underground Storage of Liquid Radioactive Wastes in the USSR, Scientific Basis for Nuclear Waste Management, Vol. 1, Ed. G.J. McCarthy, Plenum Press, NY, p. 237.

15.  H.O. Weeren, J.G. Moore, and E.W. McDaniel (1979), Waste Disposal by Shale Fracturing at ORNL, Scientific Basis for Nuclear Waste Management, Vol. 1, Ed. G.J. McCarthy, Plenum Press, NY, p. 257.

16a. R.O. Schwoebel, Stabilization of High Level Waste in Ceramic Form, Abstract, Bull. Amer. Ceram. Soc. 54, 459 (1975).

16b. J.K. Johnstone, T.J. Headley, P.F. Hlava, and F.V. Stahl (1979), Characterization of a Titanate-Based Ceramic for High Level Nuclear Waste Solidification, Scientific Basis for Nuclear Waste Management, Vol. 1, Ed. G.J. McCarthy, Plenum Press, NY, p. 211.

16c. R.G. Dosch (1979), Ceramic Forms for Nuclear Waste, ACS Symposium Series, No. 100, Radioactive Waste in Geologic Storage, Ed. S. Fried, p. 129.

17.  S. Forberg, T. Westermark, H. Larker, and B. Widell (1979), Synthetic Rutile Microencapsulation: A Radioactive Waste Solidification System Resulting in an Extremely Stable Product, Scientific Basis for Nuclear Waste Management, Vol. 1, Ed. G.J. McCarthy, Plenum Press, NY, p. 201.

18a. Committee on Ceramic Processing, National Materials Advisory Board (1968), Materials Characterization, Bulletin No. 229-M, National Academy of Science, Washington, DC.

18b. Committee on Ceramic Processing, National Materials Advisory Board (1968), Ceramic Processing, Bulletin No. 1576, National Academy of Science, Washington, DC.

19.  N.H. Macmillan, R. Roy, and P.T.B. Shaffer, Metal Matrix Radioactive Waste Forms Prepared by Electrochemical/Gravity Deposition, Nucl. Tech. 51, 97 (1980).

20.  D.J. Cameron and G.G. Strathdee (1979), Materials Aspects of Nuclear Waste Disposal in Canada, Ceramics in Nuclear Waste Management, Ed. T.D. Chikalla and J.E. Mendel, CONF-790420, Department of Energy publication, pp. 4-8.

21. K.D.B. Johnson (1979), The U.K. Program--Glasses and Ceramics for Immobilization of Radioactive Wastes for Disposal, Ceramics in Nuclear Waste Management, Ed. T.D. Chikalla and J.E. Mendel, CONF-790420, Department of Energy publication, pp. 17-24.

22. C. Sombret, R. Bonniaud, and A. Jouan (1978), Large Scale Waste Glass Production, Proc. Conf. High-Level Radioactive Solid Waste Forms, Ed. L.A. Casey, U.S. Nuclear Regulatory Commussion, Denver, CO, pp. 155-194.

23. A.S. Polyakov (1979), Principal Trends of Investigations Carried on in the USSR on Incorporation of Highly Active Wastes into Glass and Ceramic Type Materials, Ceramics in Nuclear Waste Management, Ed. T.D. Chikalla and J.E. Mendel, CONF-790420, Department of Energy publication, pp. 25-26.

24a. J. van Geel et al., Solidification of High Level Liquid Wastes to Phosphate Glass-metal Matrix Blocks, Proc. Symp. Management of Radioactive Wastes from the Nuclear Fuel Cycle, IAEA-SM 207/83, Vienna (22-26 March 1976).

24b. W. Heimerl (1979), Solidification of HLW Solutions with the PAMELA Process, Ceramics in Nuclear Waste Management, Ed. T.D. Chikalla and J.E. Mendel, CONF-790420, Department of Energy publication, pp. 97-101.

25a. M.J. Plodinec (1980), Improved Glass Compositions for Immobilization of SRP Waste, Scientific Basis for Nuclear Waste Management, Vol. 2, Ed. C.J.M. Northrup, Jr., Plenum Press, NY, p. 223.

25b. M.J. Plodinec (1979), Development of Glass Compositions for Immobilization of Savannah River Plant Waste, Scientific Basis for Nuclear Waste Management, Vol. 1, Ed. G.J. McCarthy, Plenum Press, NY, 31-35.

26a. G.J. McCarthy and R. Roy, Gel Route to Homogeneous Glass Preparation: II, Gelling and Desiccation, J. Amer. Ceram. Soc. 54, 639 (1971).

26b. G.C. Kuczynski and A.D. Silva, Formation of Glasses by Sintering, J. Amer. Ceram. Soc. 54, 51 (1971).

26c. S.P. Mukherjee et al., Comparative Study of Gels and Oxide Mixtures, J. Mat. Sci. 11, 341 (1976).

26d. J.M. Pope and D.E. Harrison, Advanced Method for Making Vitreous Waste Forms, Mat. Res. Soc. Meeting, Boston 1980.

26e. J.M. Pope and D.E. Harrison, Nuclear Waste Immobilization in Alkoxide Derived Polymer Glass Waste Management '81, Ed. R.G. Post, University of Arizona Press, 1981.

27a. J.L. Buelt, C.C. Chapman, S.M. Barnes, and R.D. Dierks (1979), A Review of Continuous Ceramic-lined Melter and Associated Experience at PNL, Ceramics in Nuclear Waste Management, Eds. T.D. Chikalla and J.E. Mendel, CONF-790420, Department of Energy publication, p. 107.

27b. M.J. Kupfer (1979), Vitrification of Hanford Radioactive Defense Wastes, Ceramics in Nuclear Waste Management, Eds. T.D. Chikalla and J.E. Mendel, CONF-790420, Department of Energy publication, p. 93.

27c. G.G. Wicks (1979), Compatibility Tests of Materials for a Prototype Ceramic Melter for Defense Glass-Waste Products, Ceramics in Nuclear Waste Management, Eds. T.D. Chikalla and J.E. Mendel, CONF-790420, Department of Energy publication, p. 82.

27d. K.D.B. Johnson (1979), The U.K. Program--Glasses and Ceramics for Immobilization of Radioactive Wastes for Disposal, Ceramics in Nuclear Waste Management, Ed. T.D. Chikalla and J.E. Mendel, CONF-790420 Department of Energy publication, pp. 17-24.

27e. T.B. Morris (1979), The Influence of Heat Transfer on the Pot Vitrification Process, Ceramics in Nuclear Waste Management, Ed. T.D. Chikalla and J.E. Mendel, CONF-790420, Department of Energy publication, p 122.

27f. Y. Sousselier (1979), Use of Glasses and Ceramics in the French Waste Management Program, Ceramics in Nuclear Waste Management, Ed. T.D. Chikalla and J.E. Mendel, CONF-790420, Department of Energy publication, p. 7.

27g. R.A. Bonniaud, N.R. Jacquet Francillon, F.L. Loude, and C.G. Sombret (1979), Glasses and Materials Used in France for Management of High-Level Wastes, Ceramics in Nuclear Waste Management, Ed. T.D. Chikalla and J.E. Mendel, CONF-790420, Department of Energy publication, p. 57.

27h. M.M. Chotin, R.A. Bonniaud, A.F. Jouan, and G.E. Rabot (1979), Operational Experience of the First Industrial HLW Vitrification Plant, Ceramics in Nuclear Waste Management, Ed. T.D. Chikalla and J.E. Mendel, CONF-790420, Department of Energy publication, p. 73.

27i. W. Heimerl (1979), Solidification of HLW Solutions with the PAMELA Process, Ceramics in Nuclear Waste Management, Ed. T.D. Chikalla and J.E. Mendel, CONF-790420, Department of Energy publication, p. 97.

27j. S. Halaszovich, E. Merz, and R. Odoj (1979), Progress in Fission Product Solidification and Characterization Utilizing the Jülich FIPS Process, Ceramics in Nuclear Waste Management, Ed. T.D. Chikalla and J.E. Mendel, CONF-790420, Department of Energy publication, p. 114.

28. W.J. Weber, R.P. Turcotte, L.R. Bunnell, F.P. Roberts, and J.H. Westsik, Jr. (1979), Radiation Effects in Vitreous and Devitrified Simulated Waste Glass, Ceramics in Nuclear Waste Management, Ed. T.D. Chikalla and J.E. Mendel, CONF-790420, Department of Energy publication, pp. 294-299.

29. G.J. McCarthy, High-Level Waste Ceramics: Materials Considerations, Process Simulation, and Product Characterization, Nucl. Tech. 32, 92 (1977).

30. A.E. Ringwood, S.E. Kesson, N.G. Ware, W.O. Hibberson, and A. Major, The SYNROC Process: A Geochemical Approach to Nuclear Waste Immobilization, Geochem. J. 13, 141 (1979).

31a. G.J. McCarthy, Radioactive Waste Management: The Nuclear Waste Form, Earth and Minerals Sciences 45, 17 (1975).

31b. G.J. McCarthy, W.B. White, and D.E. Pfoertsch, Synthesis of Nuclear Waste Monazites, Ideal Actinide Hosts for Geological Disposal, Mat. Res. Bull. 13, 1239 (1978).

31c. G J. McCarthy (1976), Crystal Chemistry of the Rare Earths in Solidified High Level Nuclear Wastes, Proc. 12th Rare Earth Res. Conf., Vail, CO, C.E. Lundin, Ed., p. 665.

31d. J.G. Pepin and G.J. McCarthy, Phase Relations in the System U-Th-Ce-RE-Zr-O: I. Literature, Bull. Amer. Ceram. Soc. 57, 358 (1978).

32. W.W. Schulz and M.J. Kupfer (1976), Solidification and Storage of Hanford's High-Level Radioactive Liquid Wastes, High-Level Radioactive Waste Management, Ed. M.H. Campbell, American Chemical Society, Washington, DC, p. 54.

33. G.S. Barney (1976), Fixation of Radioactive Waste by Hydrothermal Reactions with Clays, High-Level Radioactive Waste Management, Ed. M.H. Campbell, American Chemical Society, Washington, DC, p. 108.

34. R. Roy, E.R. Vance, and W.B. White, Ceramic Phase Compositions for Defense and Other Low Radionuclide Content Wastes, Bull. Amer. Ceram. Soc. 59, 397 (1980). (See also Ref. 48a.)

35a. F. de Korosy, Colloidal Mixtures as Batches for Glass Making, Bull. Amer. Ceram. Soc. $\underline{20}$, 162 (1941).

35b. R.H. Ewell and H. Insley, Hydrothermal Synthesis of Kaolinite, etc., J. Res. NBS $\underline{15}$, 173 (1935).

36a. D.M. Roy and R. Roy, Synthesis and Stability of Minerals in the System $MgO-Al_2O_3-SiO_2-H_2O$, Amer. Mineral. $\underline{40}$, 147 (1955).

36b. R.C. DeVries and R. Roy, Phase Equilibria in the System $BaTiO_3-CaTiO_3$, J. Amer. Ceram. Soc. $\underline{38}$, 142 (1955).

36c. D.E. Rase and R. Roy, Phase Equilibria in the System $BaO-TiO_2$, J. Amer. Ceram. Soc. $\underline{38}$, 102 (1955).

37. F.A. Mumpton and R. Roy, Hydro-thermal Stability Studies of the Zircon-Thorite Group, Geochim. Cosmochim. Acta $\underline{21}$, 217 (1961).

38. W.C. Luth and C.O. Ingamells, Gel Preparation of Starting Materials for Hydrothermal Experimentation, Amer. Mineral. $\underline{50}$, 255 (1965).

39. W.J. Lackey, P.N. Angelini, D.P. Stinton, W. Bond, and J.S. Vavruska (1980), Sol-Gel Technology Applied to Glass and Crystalline Ceramics, Proc. State of Waste Disposal Technology, Mill Tailings and Risk Analysis Models, Arizona (March 10-14).

40. A.J. Majumdar and R. Roy, System $CaO-Al_2O_3-H_2O$, J. Amer. Ceram. Soc. $\underline{39}$, 434 (1957).

41. M.J. Ruthner (1973), Preparation and Sintering Characteristics of MgO, $MgO \cdot Cr_2O_3$, and $MgO \cdot Al_2O_3$, Third Round Table Meeting, Intl. Team for Studying Sintering, Herceg-Novi, Yugoslavia (September 3-8).

42. D.M. Roy, R.R. Neurgaonkar, T.P. O'Holleran, and R. Roy, Preparation of Fine Oxide Powders by Evaporative Decomposition of Solutions, Ceram. Bull. $\underline{56}$, 1023 (1977).

43a. L.L. Ames, Jr., Cation Sieve Properties of the Open Zeolites Chabazite, Mordenite, Erionite, and Clinoptilolite, Amer. Mineral. $\underline{42}$, 1120 (1961).

43b. B.W. Mercer and L.L. Ames (1978), Zeolite Ion Exchange in Radioactive and Municipal Wastewater Treatment, Natural Zeolites: Occurrence, Properties, Use, Ed. L.B. Sand and F.A. Mumpton, Pergamon Press, NY, p. 451.

44a. W. Lutze, J. Borchardt, and A.K. Dé (1979), Characterization of Glass and Glass Ceramic Nuclear Waste Forms, Scientific Basis for Nuclear Waste Management, Vol. 1, Ed. G.J. McCarthy, Plenum Press, NY, p. 69.

44b. W. Lutze (1979), Glassy and Crystalline High-Level Nuclear Waste Forms--An Attempt at Critical Evaluation, Ceramics in Nuclear waste Management, Ed. T.D. Chikalla and J.E. Mendel, CONF-790420, Department of Energy publication, p. 47.

44c. W.F. Bonner (1979), The High-Level Waste Immobilization Program: An Overview, PNL-3094, Battelle, Pacific Northwest Laboratories, Richland, WA.

44d. G.J. McCarthy (1978), Crystalline and Coated High-Level Forms, Proc. Conf. High-Level Radioactive Solid Waste Forms, Ed. L.A. Casey, U.S. Nuclear Regulatory Commission, Denver, CO, p. 623.

45. J.E. Mendel and J.L. McElroy (June 1972), Waste Solidification Program, Vol. 10, Evaluation of Solidified Waste Products, BNWL-1666, Battelle, Pacific Northwest Laboratory, Richland, WA.

46. J.M. Rusin (1979), Multibarrier Waste Forms, Part II, Characterization and Evaluation, PNL-2668-2, Battelle, Pacific Northwest Laboratory, Richland, WA.

47. J.M. Rusin, M.F. Browning, and G.J. McCarthy (1979), Development of Multibarrier Nuclear Waste Forms, Scientific Basis for Nuclear Waste Management, Vol. 1, Ed. G.J. McCarthy, Plenum Press, NY, p. 169.

48a. R. Roy, Hydrated Ceramic Waste Forms and the Absurdity of "Leach Tests," Proceedings of the International Seminar on Chemistry and Process Engg. for High-Level Liquid Waste Solidification, Jülich, 1981.

48b. G.J. McCarthy and M. Lovette, Use of Hot Pressing to Fix Solid Radioactive Wastes in Glass, Abstract, Amer. Ceram. Soc. Bull. 51, 655 (1972).

48c. G.J. McCarthy (1973), Solid State Chemical Aspects of Radioactive Waste Disposal, Final Report, National Science Foundation, Engineering Division.

48d. G.J. McCarthy, Quartz Matrix Isolation of Radioactive Wastes, J. Mat. Sci. 8, 1358 (1973).

49. H.T. Larker (1979), Densification of Calcines and Direct Containment of Spent Nuclear Fuel in Ceramics by Hot Isostatic Pressing, Ceramics in Nuclear Waste Management, Eds. T.D. Chikalla and J.E. Mendel, CONF-790420, Department of Energy publication, p. 169.

50. G.J. McCarthy, Crystalline Ceramics from Defense High-Level Wastes, Nucl. Tech. 44, 451 (1979).

51. P.E.D. Morgan, Reactive Hot Pressing of Nuclear Waste Sludges, Bull. Amer. Ceram. Soc. 59, 397 (1980).

52a. J. Van Geel, H. Eschrich, F. Dobbels, P. Favre, and H. Sterner (1978), Incorporation of Solid High Level Wastes into Metal and Non-Metal Matrices, Proc. Conf. High-Level Radioactive Solid Waste Forms, Ed. L.A. Casey, U.S. Nuclear Regulatory Commission, Denver, CO, p. 343.

52b. J. Van Geel, H. Eschrich, and E. Detilleux, Conditioning of Solid High Level Waste Products by Dispersion into Metal Matrices, AIChE Symp. Ser. 72 (154), 161 (1976).

53. W. Neumann (1979), Embedding Methods of Solidified Waste in Metal Matrices, Ceramics in Nuclear Waste Management, Eds. T.D. Chikalla and J.E. Mendel, CONF-790420, Department of Energy publication, p. 160.

54. W.S. Aaron, T.C. Quinby, and E.H. Kobisk (1979), Development of Cermets for High Level Radioactive Waste Fixation, Ceramics in Nuclear Waste Management, Eds. T.D. Chikalla and J.E. Mendel, CONF-790420, Department of Energy publication, p. 164.

55. E. Mattson (1979), Corrosion Resistance of Canisters for Final Disposal of Spent Nuclear Fuel, Scientific Basis for Nuclear Waste Management, Vol. 1, Ed. G.J. McCarthy, Plenum Press, NY, p. 271.

56. W.A. Ross, Sintering of Radioactive Wastes in a Glass Matrix, Amer. Ceram. Soc. Bull. Abstract 54 (4), 463 (1975).

57a. P.B. Macedo et al. (1979), Stability of Fixation Solids for High-Level Radioactive Wastes, Proc. Conf. High-Level Radioactive Solid Waste Forms, Ed. L.A. Casey, U.S. Nuclear Regulatory Commission, Denver, CO, p. 81.

57b. P.B. Macedo et al. (1979), Porous Glass Matrix for Encapsulating High-Level Nuclear Waste, Ceramics in Nuclear Waste Management, Eds. T.D. Chikalla and J.E. Mendel, CONF-790420, Department of Energy publication, p. 321.

58. J.H. Simmons et al. (1979), Chemical Durability of Nuclear Waste Glasses, Ceramics in Nuclear Waste Management, Eds. T.D. Chikalla and J.E. Mendel, CONF-970420, Department of Energy publication, p. 263.

59. H.O. Weeren, Waste Disposal by Shale Fracturing at ORNL, Nucl. Eng. Design 44, 291 (1977).

60. J.G. Moore, H.E. Devaney, H.W. Godbee, M.T. Morgan, G.C. Rogers, C. Williams, and E. Newman (1978), Fixation of Radioactive Wastes in Concretes and Cementitious Grouts, Paper presented at NUCLEX, Section B-4.

61. P. Colombo and R.M. Neilson, Jr., Some Techniques for the Solidification of Radioactive Wastes in Concrete, Nucl. Tech. 32, 30 (1977b).

62. J.A. Stone (1977), Evaluation of Concrete as a Matrix for Solidification of Savannah River Plant Waste, DP-1488, Savannah River Laboratory, Aiken, SC.

63. D.M. Roy and G.R. Gouda, High Level Radioactive Waste Incorporation into (Special) Cements, Nucl. Tech. 40, 214 (1978).

64a. D.M. Roy, B.E. Scheetz, M.W. Grutzeck, A.K. Sarkar, and S.D. Atkinson (1979a), A Low Temperature Ceramic Radioactive Waste Form, Ceramics in Nuclear Waste Management, Ed. T.D. Chikalla and J.E. Mendel, CONF-790420, Department of Energy publication, p. 136.

64b. D.M. Roy, B.E. Scheetz, L.D. Wakeley, and S.D. Atkinson (1979b), Low Temperature Ceramic Waste Form; Characterization of Monazite-Cement Composites, Scientific Basis for Nuclear Waste Management, Vol. 2, Ed. C.J.M. Northrup, Jr., Plenum Press, NY.

65a. J.G. Moore, G.C. Rogers, J.H. Pachler, and H.E. Deveney (1979), FUETAP (Formed Under Elevated Temperatures and Pressures) Concretes as Hosts for Radioactive Waste, Ceramics in Nuclear Waste Management, Ed. T.D. Chikalla and J.E. Mendel, CONF-790420, Department of Energy publication, p. 132.

65b. J.G. Moore, G.C. Rogers, L.R. Dole, J.H. Kessler, M.T. Morgan and H.E. Devaney, FUETAP Concrete: An Alternative Radioactive Waste Host, Proceedings of the International Seminar on Chemistry and Process Engg. for High-Level Liquid Waste Solidification, Jülich, 1981.

66. D.M. Roy and G.R. Gouda (1974), Hot Pressed Cement in Radioactive Waste Management, Final Report to Nuclear Waste Technology Program, Subcontract BSA-481, Battelle, Pacific Northwest Laboratories, Richland, WA.

67. Commission of the European Communities, The Committee's R&D Programme (1977), Radioactive Waste Management and Storage, First Annual Progress Report, Directorate-General, Research, Science and Education, Brussels EUR5749e,f.

68. D.M. Roy (1978), Cement Based Nuclear Waste Solidification Forms, Proc. Conf. High-Level Radioactive Solid Waste Forms, Ed. L.A. Casey, U.S. Nuclear Regulatory Commission, Denver, CO, p. 479.

69. J.R. Berreth, H.S. Cole, E.G. Samsel, and L.C. Lewis (1976), Status Report: Development and Evaluation of Alternative Treatment Methods for Commercial and ICPP High-Level Solidified Wastes, ICP-1089, Idaho Falls, ID: Allied Chemical Corp., TID-4500, R64.

70a. R.H. Burns, Solidification of Low and Intermediate-Level Wastes, Atomic Energy Review 9, 547 (1971).

70b. USERDA (1977), Draft Environmental Statement, Management of Intermediate Level Radioactive Waste, ERDA-1553-D, Oak Ridge National Laboratory, Oak Ridge, TN.

# Chapter 7. STABLE MINERALS, MODELS FOR CERAMICS, APPLICATION TO PARTITIONED WASTES

by G.J. McCarthy, W.B. White, D.K. Smith and A.C. Lasaga

## Synopsis

*The role of specially selected mineral phases and assemblages for the possible case when waste streams are <u>partitioned</u> into families of radionuclides is treated in detail.*

*The predictions of thermodynamics and the evidence from nature provide incontrovertible scientific proof that even at very modest near surface temperatures (∿100°C) in the presence of water only crystalline waste forms will survive. Some crystalline minerals are formed in nature which can or do contain the (radio) nuclides of concern. Among these some are known to resist reaction in various natural rock + water environments at temperatures of a few hundred degrees centigrade. Such especially stable minerals are the obvious models for the crystalline host radiophases. This reasoning provided the theoretical base for the tailored ceramic waste forms designed by McCarthy and Roy in the early seventies and for other ceramic compositions developed since.*

*In this chapter we examine the range of mineral phases available as models. Detailed lists describing the mineralogical data on each family or phase are provided in Appendix II.*

## Introduction to Concepts

Throughout the discussions in the chapters above we have referred to one advantage of the ceramic waste forms in their being modelled on mineral assemblages which have been shown in the great experiment of geological evolution to have survived conditions not too dissimilar to the failure-case scenarios for a waste repository. Furthermore, we have made the case that waste package-rock interaction at modest 'p' and 't' will itself create new more stable (less soluble) mineral phases which will again immobilize the radionuclides in solids rather than permit their migration in fluids. Such "reactions" of waste ion and other minerals can be vastly more retarding of ion flow than the so-called "$K_d$" of the adsorption-desorption equilibria on clays and mineral surfaces.

In this chapter we bring together a catalogue of those minerals (and their properties) which show the special property that they have resisted (relatively) weathering in near surface environments. In addition we collect the data on the mineral phases which have incorporated by normal geochemical processes the natural radionuclides. Among these we also catalogue those which are known to show *crystal structural* degradation or metamictization. *One of the most important general learnings from this record is that even total metamictization (= being made noncrystalline to x-rays) hardly changes the "leach rates" of ions in the structure.* While experienced physical chemists might have predicted this on the basis of the minor differences in lattice energies of ordered and disordered states, the quantitative data in this area can serve as a useful guideline.

Lastly, we note that the use of ideal ceramic waste forms (= modelled on the most insoluble phase of that single radionuclide) is especially relevant to the case of partitioned wastes. *By partitioned waste we refer to the separation out of individual radionuclides and/or sets of radionuclides from the normal mixture in fission product streams.* This situation is not some far off future scenario. It is in fact partially accomplished in the U.S.; *perhaps half of all radionuclides in reprocessed wastes in the U.S. have been significantly partitioned* (e.g., the Cs and Sr at Hanford and more recently at Savannah River, the TRU wastes, etc.). For each partitioned waste, the goal is to create the single most stable mineral-modelled ceramic. It is in this situation that the ceramic solution is at its best, since each separate ion can be lodged in the maximally insoluble phase. (When the forty or so elements in fission products have to be simultaneously incorporated in some 6-8 mineral phases, trade-offs are unavoidable.)

### Scientific Foundation: mineral model concept

The concept of using minerals as models for the design of ceramics, of course, assumes that the reprocessing option (1) is exercised. Many of these concepts have already been utilized and described by us during the last decade (2-5) in creating tailored ceramics. The current state of our understanding of the scientific foundation for the great stability and survivability of certain minerals is discussed later in this chapter. A review of minerals that could be candidates for radionuclide fixation is also included. Only a brief overview of the scientific basis of this option is presented here.

It is a well-known fact, which is obvious from the very existence of beach sands, that there are minerals that have survived in near-surface weathering conditions for times far greater than the decay times of even the long-lived radioactive waste species. In some cases, radiochemical dating has determined ages of one to two billion years since formation for just some beach minerals. An especially relevant example of such a mineral is monazite (which has been an important component of supercalcine-ceramic formulations). Most natural monazites contain 5 to 10 wt % of Th and U oxides, and some can contain up to 35 wt % of the naturally radioactive species (6). Monazite is one of the most chemically inert of all minerals, even in constant contact with surface water at elevated temperatures. For example monazite in beach sand in the state of Guanabara in Brazil is known to be more than $10^9$ years old (7). This monazite was originally formed in pegmatites and augengneiss that at some time eroded to form a beach deposit. This deposit later became a sandstone that was still later metamorphosed at moderate temperatures with water present to form a quartzite. The current monazite beach sands were derived from a second generation of weathering of this quartzite (7). It should be noted that a few glasses are not without a demonstration of persistence *under very special conditions* in nature's laboratory, as typified by tektites, obsidian, and other related materials. Tektites are found on the surface of the earth, including the sea, as fine particles tens of microns in diameter on up to pieces about 10cm in diameter. However, these samples while very old (up to $10^8$ years) have never been heated above the surface ambients. This is a key distinction between crystalline mineral survival and natural glass survival in nature; all glasses being metastable convert to crystalline phases in the presence of water at very modest temperatures ($\sim 100°C$).

The radical distinction between glass and crystalline survivability is most evident in the fact that the total volume of glasses in the earth's crust is immeasurably small. While it is true that some archeological specimens and some lechatelierites and tektites do survive, it is to be noted that the enemies of glass survival are a wet-and-warm environment, precisely what is expected is a failed repository. The archaelogical evidence, although often cited as evidence for the possible survival of glass, in fact shows that even with considerable insulation from the ravages of nature, precious Egyptian glass art objects have frequently been corroded or transformed substantially.

The sciences of geochemistry and mineralogy can provide a sound base of knowledge about what happens in the near-surface region of the earth for millions-even billions--of years. Unlike the geophysical events that could affect the geologic and hydrologic integrity of repositories (tectonism, volcanism, etc.) that are based on probabilities, physiochemical bases for insolubility or inertness of synthetic minerals and their equilibrium vis-a-vis repository rocks can come from surveys of nature's laboratory for high-survivability minerals. Furthermore, the inertness of synthetic radioactive waste minerals can be confirmed with laboratory measurements of their interactions with water under the full range of ordinary and extra-ordinary physical and chemical repository conditions. (This is especially important for synthetic minerals containing elements such as Pu or Am.) The techniques of experimental geochemistry necessary to carry out these tests have been developed and utilized for the last four decades, and the theoretical basis for predictability is fully established in the classical thermodynamics of heterogeneous equilibria.

There are some potentially hazardous elements in nuclear wastes for which *nature* offers but little guidance on insoluble, high-stability, minerals (see later). Iodine is one such example. And, of course, the selection of candidate mineral-like hosts for manmade elements (Tc, Pm, Np, Pu, Am, Cm) would depend on crystal chemical analogy to known minerals. A modest laboratory effort would be needed to select and synthesize the inorganic mineral-like phases and verify their stability under relevant repository ambients. Fortunately, there exists for each of these man-made elements, a rather extensive crystal chemical data base of mineral-like compounds and their synthesis, gathered over the last three decades at nuclear research establishments and universities around the world.

On a worldwide basis, the most intensively studied solid for geologic disposal of reprocessing high-level wastes is glass. Because it is, by definition, a metastable solid, it can never be in equilibrium with any repository environment. Ongoing measurements of low temperature (25-70°C) radionuclide release rates of waste glasses (8,9) give cause for optimism that, within the total system of radioactive waste isolation, glass could be an adequate waste form for some geologic repository scenarios. The prominent exception is the case where water might come in contact with glass during the first few decades of storage (10). The combination of decay heat, pressure and water would create hydrothermal conditions. Here the glass could alter readily, with some elements being taken into solution and the rest being incorporated (along with components from containment and the host rock) into more stable crystalline mineral-like phases (10). The consequences of such an event for the total isolation of radionuclides may be negligible; indeed, the waste-rock interaction products could be more inert fixation hosts than was the original glass (10). But the fact remains that glass could not remain intact under these conditions and the rate of radionuclide release in the repository would be dependent on the properties of the interaction products rather than on the manmade waste solid (10,11).*

The rate of hydrothermal alteration of glass is highly sensitive to temperature, so that if glass were chosen as the waste solid, steps could be taken: (a) to protect the glass from contact with water during the first few hundred years of storage, or (b) to dilute the waste concentration, or (c) to engineer the glass composition or repository configuration so as to minimize the temperature rise after emplacement.

## Partitioned Wastes

As pointed out in a recent National Research Council report (1), glass may be an acceptable first-generation waste form, but it is certainly not the optimum form from a geochemical standpoint. With the possible exception of concerns for radiation and transmutation effects (discussed later), *assemblages of synthetic minerals appear to be as close to optimum as humans can achieve*. However, as the chapter on ceramic waste forms has shown, tailoring the mineral assemblage for all forty radionuclides leads to many compromises. The question therefore arises what mineral models would be the very best for the partitioned wastes with separated radionuclides. So

---

*Tests in Reference 10 were made at 300°C and 300 atm. These conditions *may* be somewhat more extreme than those that would exist in a wet repository, but are well within repositories still being considered worldwide.

far no estimate has been made of the degree of partitioning, the fission product concentration, or the cost of producing synthetic minerals which would exhibit significantly higher stability than glass. Until this has been done and the systems risk calculations made (which include the secondary waste streams from partitioning), it is impossible to identify the optimum system. It is clear, though, that with properly selected and tested synthetic minerals as the waste form, the concern for thermally accelerated interaction with water in a geologic repository could be eliminated. The stability and insolubility of the synthetic minerals would lead to predictably lower radionuclide release rates that could, in turn, permit less dependence on the ability of the geologic formation to retard radionuclide migration. Even if it were shown that the geology alone were sufficient to guarantee isolation for the required time period, the addition of a high-stability waste form would provide an extra margin of safety, i.e., redundant protection.

The obvious questions are: Is it feasible? What would it cost? Is it worth it? The discussion that follows will address, conceptually and necessarily incompletely, the first two questions. The third is a very complex question, involving both technical and nontechnical factors, and is clearly beyond the scope of *this* book.

Analyses of the potential hazard from certain HLLW radionuclides suggest the greatest effort in solidification into synthetic minerals would be placed on the following *groups* of elements obtained by partial partitioning:

- Actinides (An's) and Lanthanides (Ln's). The actinides Np, Pu, Am, Cm and their daughters constitute the major hazard in nuclear wastes after about $10^3$ years of storage (12). The majority of the lanthanide elements (La, Ce, Pr, Nd, Pm, Sm, Eu, Gd, Tb, Dy, Ho) are present as stable isotopes after a few years, and only trace amounts of a few Sm and Eu radionuclides have long half-lives. However, the Ln's are included with the An's for several reasons: they occur together at one stage of partitioning; Ln's and An's are crystal and chemically very similar and usually occur together in the same minerals; the Ln's can act as diluents in synthetic minerals for $\alpha$-emitting An's in order to minimize radiation effects.

- Strontium and cesium. These elements constitute both the major heat producers and biohazards in nuclear wastes for the first 600 years or so (12).

- Technetium and iodine. These two fission products have long-lived isotopes ($^{99}$Tc, $t_{\frac{1}{2}}$ = 2.1x10$^5$y; $^{129}$I, $t_{\frac{1}{2}}$ = 1.7x10$^7$y) and are biohazards. They have the additional characteristic of forming anions that can migrate in soils and rocks as fast as the solutions in which they are dissolved (13), i.e., without any substantial hold-up due to ion exchange or adsorption.

## Conceptual Processes and Products

The various conceptual flow sheets for this option are summarized in Figure IX. Dissolution of spent fuel would release the first waste product, iodine, along with the other volatiles. High-level liquid waste results from the Purex process in which 99 to 99.9% of the U of U+Pu are

recovered from the spent fuel. One option would be to solidify the HLLW
as a complex assemblage of synthetic minerals designed to be in equilibrium with a particular geologic rock type. This would utilize the PSU/PNL
supercalcine-ceramic concepts and processing. Alternatively, the HLLW
could be partitioned into several waste streams followed by solidification
of each stream into optimized synthetic minerals selected for their
stability in a particular geologic repository.

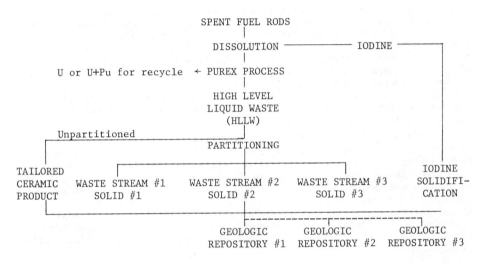

Fig. IX. *Showing relation of unpartitioned and partitioned approaches to
solidification.*

*Tailored Ceramics.* This waste form is defined as an essentially
crystalline assemblage of mutually compatible, refractory and leach-
resistant phases made by modifying the composition of HLLW with selected
liquid additives prior to calcining (3,4). The supercalcine powder is
processed to a monolithic ceramic by pelletizing and sintering or by hot
pressing. Research on this product has been underway at PSU since 1973.
Four engineering-scale demonstrations of supercalcine processing using
chemically simulated (that is nonradioactive) wastes have been performed at
PNL. The supercalcine powders have also been consolidated into spherical
ceramic cores for use in a multibarrier waste form (14). Thus the
multiphase ceramic option is an engineered alternative today.

Although high-stability minerals have been used as guides for selection
of ceramic phase assemblages, these products have not been purposely designed
to be in equilibrium with any particular geologic rock type. Nevertheless,
many of the synthetic minerals in existing supercalcine-ceramic products
should be stable in contact with igneous rocks (e.g., granite, basalt)
because they are analogous to certain accessory minerals commonly found in
these rocks. Some of the structure types and compositions (4,14) of these
potentially stable alumina or titania-rich ceramic phases have been listed
in Tables 10 and 11.

Research at The Pennsylvania State University has shown that the stability of designed ceramic products in basalts and shales, in repository-simulating tests are substantially better than that of glass. However, the net radionuclide release from the total waste package may not depend crucially on the waste form alone (see Chapter 10).

It would be a clear advantage to be able to solidify the complete HLLW stream into a suitable synthetic mineral assemblage. However, the requirement for mutual compatibility (4) among the many phases in supercalcine-ceramics could make it a challenge to design synthetic minerals for An's, Ln's, Sr, Cs and Tc that are simultaneously compatible among themselves and with a selected geologic repository host rock. Partitioning, that is, separating HLLW into two or more fractions, would of course simplify the task, give a higher probability of early success and open up the possibility of using multiple repository sites selected for stability with the various synthetic mineral assemblages made from each fraction. As has been pointed out, each partitioning step increases costs and occupational exposures and creates secondary waste streams. All these factors must be considered in the final analysis of the system.

*Partitioning Possibilities.* Because of the great complexity of HLLW there are numerous choices for partitioning. Nine waste streams arising from various partitioning possibilities, plus iodine collection from the dissolver off-gas, are illustrated in Figure X. These have been identified through considerations of crystal chemistry and stability of candidate synthetic minerals, the three groups of radionuclides of concern (see above) and the known techniques for isolating various waste fractions. The following is a qualitative, and primarily conceptual, discussion of the incorporation of the nine waste streams into the synthetic mineral/geologic disposal option.

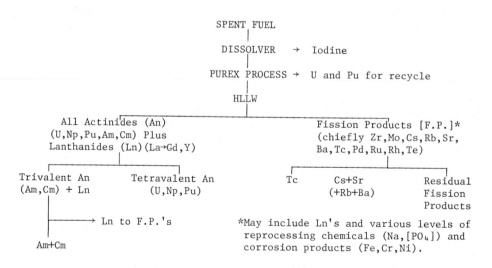

Fig. X. *Various levels of partitioning.*

*Actinide Plus Lanthanides*. Separation of the actinides plus lanthanides from the balance of the HLLW species involve an additional step beyond the Purex process. It would remove much of the long-term risk from the residual HLLW fraction. The mixture of An's and Ln's could then be mixed with appropriate additives and prepared as one or more optimum synthetic minerals. For example, it is likely that a single monazite-structure ($LnPO_4$) solid solution phase could be prepared simply by adding appropriate phosphate and charge balancing cations (e.g., $Ca^{2+}$ to Balance $An^{4+}$) to the liquid (An+Ln) waste stream and calcining. After crystallization and consolidation, the synthetic monazite could be packaged and shipped to, say, a granite geologic repository. Monazite, a highly insoluble mineral in its own right, should be in or near equilibrium with granitic mineral assemblages over the full range of repository ambients. Solubility-transport experiments and modeling should therefore be able to confirm very minimal release of actinides and their daughters even in the event of catastrophic failure of the repository. This is just one example of an (An+Ln) solidification-geologic disposal scenario. In any case, the greatest care in selection of the synthetic minerals(s) and the geologic repository would want to be taken with this waste fraction.

A complication, typical of all partitioning-based disposal options, should be noted. According to projections based on *existing* techniques, a small amount (<2 wt %) of fission products (Zr,Tc,Ru,Pd) occur in the An+Ln stream. Consideration would have to be given to whether these would crystallize in the mozazite solid solution (Zr almost surely would) or as one or more separate minor phases. It is possible that a second synthetic mineral phase of Tc would be required. The Ru and Pd might crystallize as oxides.

*Fission Products*. Application of known partitioning techniques would only remove 95 to perhaps 99.9% of the residual An's in HLLW, and there is a problem with continuous verification of the level of An removal. Future technologies may increase the efficiency of actinide removal without unduly increasing the costs, but it is unlikely that partitioning to give an "An-free" HLLW, along with the required verification, will be feasible in the near future. What one would gain through the extra expense of (Ln+An) partitioning would be a less hazardous fission product waste fraction, especially after the first $10^3$ years or so of storage. The gain in long-term risk will be a function of the actinide removal efficiency. Whatever the gain, it must be balanced against the fiscal and environmental cost of multiple waste streams and repositories. Modeling of the risk inherent in this waste fraction could indicate that long-term (>$10^3$ year) requirements on the stability of *its* geologic repository formation could be relaxed. The bulk of the heat-producing radionuclides would remain with this fraction, so attention would still have to be given to the stability of the waste form and the near-field thermal effects on the repository. Incidentally, the volume of this fraction could be somewhat smaller due to removal of the Ln's (which typically make up 20-30 wt % of HLW oxides) along with the An's.

Partitioning into the two waste streams just described appears to be feasible, based on laboratory-scale experiments,* and could probably be

---

*With bidentate extraction technology under development at the Oak Ridge National Laboratory.

fully demonstrated in the near term (approximately 5-10 years). It appears to be a relatively "clean" process in that only small amounts of additional processing wastes are generated. These wastes, primarily Na and phosphate, would be added to the fission product fraction. However, in five of the six partitioning waste streams to be described below (Cs+Sr is the exception), separations technology development is either at a very early stage or merely conceptual). The effectiveness of the partitioning is often much lower than with (An+Ln) separation from fission products, and large amounts of processing wastes result from the multistep operations. Therefore, it must be emphasized that any application of the synthetic minerals concept to *these* waste streams would require prior or concurrent development of partitioning technologies.

*Cesium plus Strontium*. Removal of these elements, the major heat producers among the fission products, would virtually eliminate concern for any thermally induced problems with the waste form and geologic repository chosen for the residual fission product fraction. Stable Rb and Ba would also be included in this fraction because of their chemical similarity to Cs and Sr. The Cs+Sr (plus Rb and Ba) fraction could be solidified into one or more synthetic minerals that would be stable in the chosen geologic repository even under hydrothermal conditions at hundreds of degrees above ambient (10). An assemblage of Cs-Rb-Na-Sr-Ba-aluminosilicates is one promising candidate.

Cesium and strontium separation from defense high-level wastes has recently been completed at the Hanford Reservation (15). At the Savannah River Plant the reference process removes Cs but does not recover it as a distinct product (16). Note that relatively large amounts of sodium process wastes are produced in the existing processes. The presence of this sodium would act to dilute the radionuclide Cs and Sr and reduce the heat output of the bulk synthetic mineral product. Natural non-radioactive Sr and Cs could also be used as a diluent if this proved to be necessary. Application of the existing technology would remove only 60-80% of the (Cs+Sr) from the residual fission products mix. This level of separation would go a long way toward lowering the decay heat output of this mix, but it would certainly not eliminate its short-term $^{90}$Sr and $^{137}$Cs hazards. Vitrification of the residual fission products could be attractive because of the glass product would be largely free of any concerns about devitrification or hydrothermal alteration (10) during the "thermal period" of geologic disposal. Alternatively, the supercalcine-ceramic concepts and technology could be readily applied to synthesize a crystalline mineral-like assemblage from this mix.

*Technetium*. The long-term radiological risk of the fission product mix could be further reduced by removal of $^{99}$Tc. There are, of course, no technetium minerals. In crystal chemical investigations (17-19), Tc has been found to crystallize in inorganic compounds isostructural with the minerals spinel, scheelite and perovskite, so these phases might be possible hosts. Rhenium minerals and their natural stabilities might also suggest candidates for a Tc host. The problems of entrainment of Tc in the dissolver solids and the additional complexity of the Tc partitioning are discussed later. Obviously, considerable research on Tc separations, synthetic mineral preparation, and stability in geologic repositories would need to be performed before a comprehensive assessment of costs and benefits of Tc separations could be carried out.

*Separations Within the Actinides and Lanthanides*. Three other
partitioning possibilities are shown in Figure X. Two waste fractions
could be created by separation of the "trivalent" An's (Am and Cm) plus the
Ln's from the "tetravalent" An's (U,Np,Pu). With these separate streams
the possibility of even more effective design of stable synthetic minerals
would be open. For example, the (Ln's+Am+Cm) could be crystallized as a
monazite and (U+Pu+Np) formed into a fluorite structure solid solution
$(U,Np,Pu)O_2$. The latter phase could be diluted, if necessary to reduce
solubility or radiation effects, with isostructural $CeO_2$. Alternatively,
it might be desirable to recycle the U, Np and Pu for use in isotope
production (e.g., $^{237}Np$ for $^{238}Pu$ production) or nuclear fuel applications.

Yet another partitioning possibility would be to separate Am and Cm
from the far more abundant, but stable, Ln's. The potential advantage
might come in reducing the volume of the most hazardous wastes to which the
greatest care in solidification and geologic disposal would have to be
applied. A potential disadvantage could be the high concentrations of
$\alpha$-emitters.

Clearly, there would have to be very strong motivation provided by
rich-benefit analyses to justify the extra cost, complexity and generation
of secondary wastes accompanying additional separations within the (An+Ln)
waste stream.

*Iodine*. This element is separated from commercial nuclear wastes
during dissolution of the spent fuel rather than by partitioning of HLLW.
Nature offers little guidance on insoluble, high-stability minerals.
Virtually all iodine minerals are water soluble and are found in evaporite
deposits. The best synthetic mineral discovered to date is NaI-sodalite
$(Na_8Al_6Si_6O_{24}I_2)$. Alternatively, iodine could be solidified by the barium
iodate-cement or lead-exchanged zeolite processes under development at the
Oak Ridge National Laboratory (20) and the Idaho National Engineering
Laboratories (21).

*Ceramic Radiophases for Partitioned Wastes*. A detailed discussion of
synthetic mineral preparation techniques has been given in Chapter 5.
Briefly, these are:

- calcination: mixing of waste and mineral-forming additive *liquids*
  followed by a rapid drying-calcination step and an additional heat
  treatment to crystallize and consolidate the synthetic mineral; an
  alternative is to interpose the step of making a gel.

- solid state reaction: mixing of calcined waste and mineral-forming
  additives as powders followed by a crystallization-consolidation step;
  gel-crystal reactions are accomplished at lower temperatures and/or
  time periods.

- hydrothermal synthesis: a batch process in which waste and mineral-
  forming additives would be reacted in a pressure vessel which would
  also serve as the final canister at elevated temperatures with water
  acting as both solvent and catalyst.

In terms of simplicity and ease of adaption to remote conditions, the order
of preference of the three methods is hydrothermal in-can reaction >
calcination > solid state reaction.

The first method should be applicable to all of the waste streams just described because they occur as liquids at the end of the partitioning (or in the case of I, the scribbing of dissolver off-gas) step. If the required synthetic minerals could be formed by the calcination technique (and it is likely that most would), then the spray or fluidized bed calcination technologies developed at many laboratories could be readily applied. For several years the supercalcine-ceramic synthetic mineral assemblages (4) have been routinely prepared (in non-radioactive simulations) with an engineering-scale spray calciner, disc pelletizer and resistance heated furnace at the Pacific Northwest Laboratories (14). In the reference synthetic minerals/geologic disposal scenario described below, the calcination or gel route is assumed for each of the waste streams.

TABLE 16

Synthetic Mineral Assemblages and Repositories in the Reference Scenario

| WASTE FRACTION | SYNTHETIC MINERALS ASSEMBLAGE | REQUIRED ADDITIVES | REPOSITORY ROCK |
|---|---|---|---|
| An + Ln | $(Ln,An,Ca)PO_4$ MONAZITE | $H_3PO_4$ $Ca(NO_3)_2$ | GRANITE, SHALE OR BASALT |
| Cs + Sr (+Na+Rb+Ba) | $(Cs,Na,Rb)AlSi_2O_6$ Na-POLLUCITE $(Ba,Sr)_{1-x}Na_xAl_{2+x}Si_{2+x}O_8$ FELDSPAR | $Al(NO_3)_3$ LUDOX-AS (colloidal $SiO_2$) | GRANITE, SHALE OR BASALT |
| RESIDUAL FISSION PRODUCTS, PROCESSING CHEMICALS AND CORROSION PRODUCTS (Ar,Mo,Na,Fe, [PO₄],Ru,Tc,Pd, Ni,Cr,Rh,Te) | $CaNa_6Al_6Si_6O_{24}(MoO_4)$ Mo-SODALITE $ZrO_2$ $(Fe,Cr)_2O_3$ BADDELEYITE HEMATITE $Ca_5(PO_4)_3OH$ $(Ni,Fe)(Fe,Tc,Cr,Rh)_2O_4$ APATITE TREVORITE $RuO_2$ PdO $Ca_3TeO_6$ $NaAlSiO_4$ NEPHELINE | $Ca(NO_3)_2$ $Al(NO_3)_3$ LUDOX-AS | ANY ROCK |
| I | $Na_8Al_6Si_6O_{24}I_2$ I-SODALITE | $NaNO_3$ $Al(NO_3)_3$ LUDOX-AS | SALT |

A combined (An+Ln) fraction would be partitioned as a nitrate solution from the fission products and would be combined with phosphoric acid (with some Ca [as $Ca(NO_3)_2$] to charge balance $An^{4+}$) followed by calcining, pelletizing, crystallization-consolidation and packaging in a metal canister for shipment to and emplacement in a granite or basalt repository.

The high-heat elements, Cs and Sr, would be separated from the fission products, mixed together, and, again using the reference process, would be converted to a mixture of Na-pollucite and Ba-Sr-plagioclase feldspar pellets for packaging, shipment and disposal in a granite or basalt repository. If the thermal loading from the product synthesized directly from the waste stream were too great for the mechanical integrity of the granite repository, then the radionuclides could be diluted into solid solution with isostructural stable (i,e., non-radioactive) minerals. For example, if the loading of ($^{90}Sr + {}^{137}Cs$) were 15 wt % and this proved to generate too much heat, then the additives to the waste stream could include sufficient components of "cold" pollucite and "cold" feldspar to bring the loading down to an acceptable 5 or 10 wt % while still maintaining the same mineralogy.

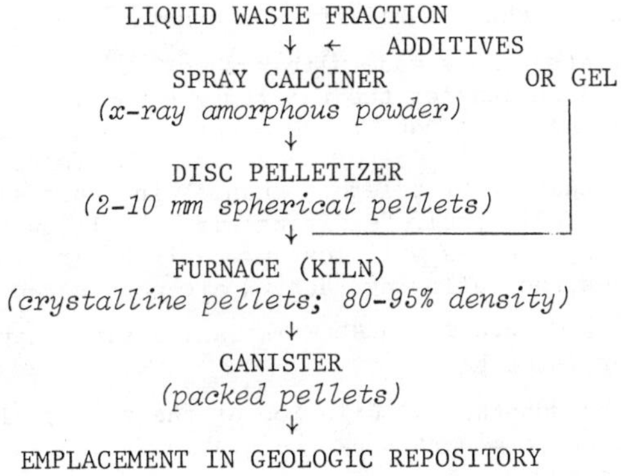

LIQUID WASTE FRACTION
↓ ← ADDITIVES
SPRAY CALCINER          OR GEL
*(x-ray amorphous powder)*
↓
DISC PELLETIZER
*(2-10 mm spherical pellets)*
↓ ——————————
FURNACE (KILN)
*(crystalline pellets; 80-95% density)*
↓
CANISTER
*(packed pellets)*
↓
EMPLACEMENT IN GEOLOGIC REPOSITORY

*Fig. XI.  Reference Process for Each Waste Fraction.  (The nature of the product after each processing step is given in italics.)*

The synthetic mineral assemblage of the residual fission product waste stream was conceived generally according to supercalcine-ceramic phase formation models (4,14). The likely hosts for any unseparated Cs would be the sodalite and/or nepheline phases, and for Si, the sodalite, apatite and/or $Ca_3TeO_6$ phases. The reasoning for including any rock (including salt) which meets general repository requirements is the vast reduction in heat generation and radiotoxicity of this fraction. However, the conceptual phase assemblage would be closest to equilibrium in an aluminosilicate (e.g., granite, basalt, shale) repository environment.

In this reference scenario, the HI waste stream would be processed into a synthetic I-sodalite and sent to a salt repository.

*Radiation and Transmutation Effects*. It has been recognized for some time that crystalline ceramic (= synthetic mineral) waste form phases that contain significant amounts of radionuclides could be susceptible to variations in physical (e.g., density) and chemical (e.g., solubility) properties due to the effects of radioactive decay (4,23). These effects can be separated into two classes:

- Metamictization: the loss of crystallinity in certain minerals and ceramics due to $\alpha$-radiation bombardment from contained An's.

- Transmutation effects: $\beta$-decay processes such as $^{137}Cs \rightarrow ^{137}Ba$ or $^{90}Sr \rightarrow ^{90}Y \rightarrow ^{90}Zr$ could cause substantial structural changes on an atomic scale and would release electrons to the structure. Accompanying $\gamma$-emission results in ionization effects.

Nature's laboratory provides some insight into the process and products of metamictization (see Chapter 8 and Appendix II). Some An (Th and U)-containing minerals are *never* metamict while others often occur with at least some metamictization. Thus, there are apparently structural controls on susceptible to secondary alteration. Yet zircon, a mineral that is often metamict (25), is perhaps the most indigenous heavy mineral in stream beds and beach sands. This suggests that metamictization might have little effect on the solubility of synthetic mineral hosts for An's.

Several comments can be made with respect to two of the synthetic mineral assemblages chosen for the reference scenario. Monazite, the host for the $\alpha$-emitting An's, is one of the minerals that is rarely, if ever, metamict. The conceptual $^{137}Cs$-pollucite and $^{90}Sr$-feldspar hosts are both network silicates that might tolerate changes in ionic size and charge from some of the ions in their "stuffing sites." It must also be noted that the transmuting species will most probably be present in rather small concentrations when the following factors are considered:

- not all of the Cs and Sr in HLLW is radioactive—approximately 40% of the Cs and Sr is stable.

- there would be substantial dilution of the radioactive nuclides in the host synthetic minerals by the stable Na, Rb and Ba components of the waste stream.

- It is quite possible that reductions in thermal output mandated by rock mechanical considerations at the repository would require further solid solution dilution of the radionuclides; dilution could be used to minimize any deleterious transmutation effects. *Dilution by solid solution (=identical, isovalent or balanced aliovalent substitution in the lattice) is universally available in every single ceramic radiophase.*

Similar structural and compositional control could be used to minimize any deleterious effects from the decay of $^{129}I$ into Xe in the zeolite-like sodalite host mineral.

# REFERENCES FOR CHAPTER 7

1. U.S. Department of Energy, Directorate of Energy Research, Report of Task Force for Review of Nuclear Waste Management--Draft. DOE/ER-0041D, pp. 161-162, February 1978.

2. J.L. McElroy, Ed. Quarterly Progress Report--Research and Development Activities--Waste Fixation Program--July through September 1974. BNWL-1871, Battelle, Pacific Northwest Laboratories, Richland, Washington, pp. 49-57. November 1974. (See also all subsequent reports in this series.)

3. G.J. McCarthy, "High-Level Waste Ceramics." Trans. Am. Nucl. Soc. 23:168-169 (1976).

4. G.J. McCarthy, "High-Level Waste Ceramics: Materials Considerations, Process Simulation and Product Characterization." Nucl. Technol. 32: 92-105 (1977).

5. R. Roy, "Rational Molecular Engineering of Ceramic Materials," J. Am. Ceram. Soc. 60:350-363 (1977).

6. W.A. Deer, R.A. Howie and J. Zussman, Rock Forming Minerals--Vol. 5--Non-Silicates. Longman Group, Ltd., London, pp. 338-346 (1962).

7. O.H. Leonardos Jr., "Origin and Provenance of Fossil and Recent Monazite Deposits in Brazil." Economic Geology 69:1126-1128 (1974).

8. J.E. Mendel et al. Annual Report on the Characteristics of High-Level Waste Glasses. BNWL-2252, Battelle, Pacific Northwest Laboratories, Richland, Washington, 99 pp., June 1977.

9. L.D. Ramspott, Ed. Waste Isolation Projects--FY-1977. UCRL-50050-77, Lawrence Livermore Laboratory, Livermore, California, 24 pp., January 1978.

10. G.J. McCarthy, W.B. White, R. Roy, B.E. Scheetz, S. Komarneni, D.K. Smith and D.M. Roy, "Interactions Between Nuclear Waste and Surrounding Rock." Nature 273:216-217 (1978).

11. National Research Council, Committee on Radioactive Waste Management, Panel of Waste Solidification, Solidification of High-Level Radioactive Wastes (in press).

12. B.L. Cohen, "High Level Radioactive Wastes from Light Water Reactors." Rev. Modern Phys. 49:1-19 (1977).

13. D. Rai and R.J. Serne, Solid Phases and Solution Species of Different Elements in Geologic Environments. PNL-2651, Battelle, Pacific Northwest Laboratories, Richland, Washington, March 1978.

14. J.M. Rusin, R.O. Lokken, J.M. Lucas, K.R. Sump, M.F. Browning and G.J. McCarthy, Multibarrier Waste Froms--Part I: Development. PNL-2668-1, Battelle, Pacific Northwest Laboratories, Richland, Washington, 103 pp. (in press).

15. Energy Research and Development Administration, Alternatives for Long-Term Management of Defense High-Level Wastes, Hanford Reservation. ERDA 77-44, September 1977.

16. U.S. Department of Energy, Draft Environmental Impact Statement. Long-Term Management of Defense High-Level Radioactive Wastes, Savannah River Plant, DOE/EIS-0023-D, July 1978.

17. O. Muller, W.B. White and R. Roy, "Crystal Chemistry of Some Technetium-Containing Oxides." J. Inorg. Nucl. Chem. 26:2075-2086 (1964).

18. J. Wassilopulos, Polyoxides of Four and Seven Valent Technetium with Alkaline Earths. KFK-341, Kernforschungszentrum Karlsruhe, 66 pp., July 1965.

19. C. Keller and B. Kanellakopulos, "Ternary Oxides of Three to Seven Valent Technetium with Alkali Elements," J. Inorg. Nucl. Chem. 27: 787-795 (1965).

20. J.G. Moore, H.E. Devaney, M.T. Morgan and G.C. Rogers, "Fixation of $^{129}I$ in Portland Cement," (Abstract). Am. Ceram. Soc. Bull. 57:324 (1978).

21. C.M. Slansky et al. Technical Division Quarterly Progress Report, July 1-September 30, 1977. ICP-1132, Idaho National Engineering Laboratory, Idaho Falls, Idaho, October 1977.

22. J.D. Bredehoeft, A.W. England, D.B. Stewart, N.J. Trask and I.J. Winograd, Geologic Disposal of High-Level Radioactive Wastes--Earth Science Perspectives. Circular 779, U.S. Geological Survey, May 1978.

23. D.W. Readey and C.B. Cooley, Eds. Glass and Ceramic Radioactive Waste Forms. ERDA CONF-770102, Energy Research and Development Administration, Washington, DC, January 1977.

24. G.J. McCarthy and M.W. Grutzeck, Preliminary Evaluation of the Characteristics of Nuclear Wastes Relevant to Geologic Storage in Basalt. RHO-C-12, Rockwell Hanford Operations, Richland, Washington, 29 pp., May 1978.

25. A.E. Ringwood, Safe Disposal of High-Level Nuclear Reactor Wastes: A New Strategy, Australiam National University Press, Books Australian, Norwalk, CT, July 1978.

26. J.M. Rusin, J.W. Wald, "Radiation Effects." Quarterly Progress Report: Research and Development Activities Waste Fixation Program, J.L. McElroy, Ed., PNL-2265-2, Pacific Northwest Laboratory, Richland, Washington, July 1978.

# Chapter 8  RADIATION AND TRANSMUTATION EFFECTS IN WASTE FORMS

## by E.R. Vance

Synopsis

*Solids which incorporate waste radionuclides are continually exposed to a radiation field generated by the decay of the radionuclides. The radiation field includes α-recoil nuclei, α-particles, β and γ-rays, neutrons and fission fragments. Another aspect of the decay process which affects the stability of the solid is the transmutation of the original parent nuclide to the final nuclide daughters. This chapter reviews some of the available data relating to the effects of radiation and transmutation on solids in general, and then discusses these effects in various nuclear solid waste forms--glasses, ceramics, concretes, etc. Long-term predictions of the effects of radiation and transmutation can sometimes be made on the basis of observations on ancient natural minerals as well as from real-time laboratory studies. When a waste solid is in contact with aqueous solution in a geological repository, radiation will affect the aqueous solution as well as the solid, adding a new dimension to the problem of dissolution of the solid.*

*Though much work remains to be done, some general conclusions can be reached at present. For the various phases likely to be present in ceramic waste forms, there will be a wide ($\sim 10^3$) variety of responses by different phases to various kinds of radiation. Given the radiation levels required to produce major changes in candidate radiophases, there are grounds for expecting that radiation and transmutation effects in glasses, ceramics or concrete incorporating U.S. defense wastes will not be serious, simply because the radionuclides in such wastes are present in such low concentrations. In commercial wastes, the effects are $\sim 10^2$ times larger, and the aim of waste form optimization is to ensure that all the radionuclides enter relatively highly radiation-resistant crystalline phases or glasses. Mitigation of transmutation effects is much more difficult; so is their study. The effects of radiation on likely repository waters, overpacks, or canisters have not been explored in detail. Here again there are no grounds for expecting substantial problems for aged (>20 yr old) wastes.*

# Chapter 8.  RADIATION AND TRANSMUTATION EFFECTS IN WASTE FORMS

## Introduction

Solids presently under serious consideration for the permanent disposal of high-level nuclear wastes in the seabed or geological repositories would be processed mixtures of nuclear waste and inert additives such as cements, zeolites, glass-forming oxides or frits, etc. The most important properties of these solids will be the leachabilities in various groundwaters and the waste loading, that is, the fraction of the waste in the processed solid.

In principle, self-irradiation and transmutation effects *could* significantly change the properties of the waste form with time.  The general idea that radiation effects will be deleterious is based on the well-known lattice disruption and dimensional changes, and the increased rate of etching--α- and fission track development--in even refractory solids, both crystalline and vitreous.  Radiolysis effects might also be expected to be damaging because of the very reactive nature of the free radicals produced both in the solid and the leaching solution.

While this chapter deals with radiation effects in candidate nuclear waste forms, it must of course be remembered that these effects are only one parameter set entering into the design of a solid waste form.  Moreover, the properties of solid waste forms, though undeniably important, are only one parameter set in the whole operation of nuclear waste disposal.

TABLE 17

Yields of Stable (or Long-Lived)* Fission Products from Thermal Fission of $^{235}U$ (2)

| (Z) Element | $A_R$ | % Yield† | % RA | $t_{1/2}$ yr. | (Z) Element | $A_R$ | % Yield† | % RA | $t_{1/2}$ yr. |
|---|---|---|---|---|---|---|---|---|---|
| (34) Selenium | 79 | 0.4 | 11 | ~6x10⁴ | (50) Tin | 126 | 0.1 | 37 | 1x10⁵ |
| (35) Bromine | | 0.3 | -- | S | (51) Antimony | 125 | 0.1 | 30 | 2.7 |
| (36) Krypton | | 3.8 | | | (52) Tellurium | | 2.5 | -- | S |
| (37) Rubidium | | 1.3 | -- | S | (53) Iodine | 129 | 1.0 | 83 | 1.6x10⁷ |
| (38) Strontium | 90 | 6.2 | 60 | 29 | (54) Xenon | | 20 | | |
| (39) Yttrium | | 4.8 | -- | S | (55) Cesium {134 / 135 / 137} | | 20 | 5 / 15 / 44 | 2 / 3x10⁶ / 30 |
| (40) Zirconium | 93 | 36.9 | 20 | 1.5x10⁶ | (56) Barium | | 6.7 | -- | S |
| (41) Niobium | | <0.1 | -- | | (57) Lanthanum | | 6.6 | -- | S |
| (42) Molybdenum | | 25.0 | -- | S | (58) Cerium | 144 | 12.3 | 2 | 0.8 |
| (43) Technetium | 99 | 6.1 | 100 | 2x10⁵ | (59) Praseodymium | | 5.9 | -- | S |
| (44) Ruthenium | 106 | 9.3 | 5 | 1.0 | (60) Neodymium | | 20.5 | -- | S |
| (45) Rhodium | | 4.9 | -- | S | (61) Promethium | 147 | 2.3 | 100 | 2.6 |
| (46) Palladium | 107 | 1.4 | 15 | 7x10⁶ | (62) Samarium | 151 | 1.9 | 5 | 90 |
| (47) Silver | | 0.2 | -- | S | (63) Europium {154 / 155} | | 0.2 | ~28 / 2 | 85 / 49 |
| (48) Cadmium | | 0.1 | -- | S | (64) Gadolinium | | <0.1 | -- | S |
| (49) Indium | | <0.1 | -- | S | | | | | |

*$t_{1/2}$ > 0.5 yr.
†% Yield is the number of atoms per 100 fissions.
S = stable.

RA = radioactive.
$A_R$ = atomic weight of radioactive isotope.

Inventories of high-level nuclear waste in the U.S. have been presented in Tables 4 and 5. The content of long-lived radioactive nuclides in fission products is indicated in Table 17 and the chemical compositions of representative reprocessing wastes are repeated in Table 18. In contrast, the large volumes of nuclear wastes at various defense establishments are much more dilute in fission products (4-8). As an example, some compositions for defense wastes at the Savannah River plant are given in Table 19.

## TABLE 18

### Composition of High-Level Nuclear Wastes from Reprocessing

#### Mole Percent

| Component | PW-4b | PW-7 | PW-7a | PW-8a |
|-----------|-------|------|-------|-------|
| Rare earths | 26.4 | 30.9 | 20.5 | 6.7 |
| Zr | 13.2 | 9.2 | 6.1 | 3.5 |
| Mo | 12.2 | 8.2 | 5.5 | 3.2 |
| Ru | 7.6 | 5.1 | 3.4 | 2.0 |
| Cs | 7.0 | 4.7 | 3.1 | 1.8 |
| Fe | 6.4 | 8.7 | 5.8 | 30.0 |
| Pd | 4.1 | 2.8 | 1.8 | 1.1 |
| Sr | 3.5 | 2.3 | 1.6 | 0.9 |
| Ba | 3.5 | 2.3 | 1.6 | 0.9 |
| $[PO_4]$ | 3.2 | 8.7 | 5.8 | 1.7 |
| U | 1.4 | 9.6 | 6.3 | 3.7 |
| Rb | 1.3 | 0.9 | 0.6 | 0.3 |
| Na | -- | 0.9 | 34.0 | 40.0 |
| Am+Cm+Pu | 0.2 | 0.2 | 0.2 | 0.1 |
| Others | 9.8 | 5.5 | 3.7 | 4.1 |

PW is U.S. Energy Research and Development Administration shorthand for Purex process Wastes.

PW-7 contains an excess of gadolinium used as a neutron absorber plus more iron, $[PO_4]$, uranium, and some sodium.

PW-7a is typical of HLW which was intended to be generated by Allied General Nuclear Services, Barnwell, South Carolina. PW-8a is typical of HLW in storage at Nuclear Fuel Services, West Valley, New York.

TABLE 19

Approximate Compositions of Savannah River Waste (4)

### Major Elements

| Component | High Fe | Composite | High Al |
|-----------|---------|-----------|---------|
| $Fe_2O_3$ | 53.17 | 36.13 | 5.32 |
| $Al_2O_3$ | 4.89 | 28.26 | 76.05 |
| $MnO_2$ | 3.56 | 9.94 | 4.37 |
| $U_3O_8$ | 12.34 | 3.26 | 1.28 |
| CaO | 3.62 | 2.69 | 0.35 |
| NiO | 9.08 | 4.47 | 0.78 |
| $SiO_2$ | 0.40 | 0.85 | 0.56 |
| $Na_2O$ | 4.52 | 5.08 | 1.96 |
| $Na_2SO_4$ | <0.50 | 0.93 | <0.50 |
| zeolite* | 8.82 | 8.93 | 9.33 |

*Ion-Siv IE-95 zeolite; mixture of $CaAl_2Si_4O_{12} \cdot 6H_2O$
and $(Na,K,Ca)_3Al_3Si_9O_{24} \cdot 8H_2O$.

### Minor Elements

| Element | Range, Wt % | Element | Range, Wt % |
|---------|-------------|---------|-------------|
| C | 0-16.8 | Nb | 0.02-0.1 |
| Nd | 0.3-1.0 | Hg | 0.1-2.8 |
| Zr | 0.1-0.5 | Y | 0.01-0.05 |
| Cr | 0.01-0.45 | Ag | 0-0.3 |
| Ru | 0.03-0.5 | Pm | 0-0.04 |
| Ba | 0.1-0.25 | Pb | 0.1-0.5 |
| Ce | 0.05-1.0 | Ti | 0-0.1 |
| K | 0.02-0.14 | Sm | 0-0.05 |
| Cl | 0-2.0 | Eu | 0-0.02 |
| Sr | 0-0.15 | V | 0-0.01 |
| La | 0.04-0.3 | Rh | 0-0.05 |
| Pr | 0.04-0.2 | F | 0-0.15 |
| S | 0.02-0.5 | Cs | 0-0.06 |
| P | 0.03-0.3 | B | 0-0.05 |
| Zn | 0.04-0.5 | Cd | 0-0.05 |
| Th | 0-0.18 | Pb | 0-0.03 |
| Mg | 0.06-0.3 | Mo | 0-0.01 |
| Cu | 0.03-0.15 | Co | 0-0.01 |

Many of the minor elements are largely fission
products. Principal radionuclides (in mCi/gm)
are: $^{90}Sr$ (30-50), $^{144}Ce$ ($\sim$2), $^{106}Ru$ ($\sim$1),
$^{137}Cs$ ($\sim$0.5), $^{154}Eu$ ($\sim$0.3), $^{125}Sb$ ($\sim$0.4), $^{60}Co$
($\sim$0.2), gross $\alpha$ ($\sim$0.3).

Over the last 30-odd years there has been a vast amount of work
done on fundamental radiation effects in all types of solids. However,
virtually all this work has addressed effects (structural, mechanical,
etc.) other than the solubility behavior of interest in the present
context. Even though fission and $\alpha$-recoil tracks can be "developed"
by etching in some solids, the etchants are specific and invariably
highly corrosive--strong HF and NaOH solutions feature prominently.
This fact of itself is of considerable interest as it might be thought
that to demonstrate a leach enhancement of a damaged region versus a
non-damaged region, the best solvent would be one which only barely
affected the undamaged region. That this view is not supported

may mean that the solution chemistry of the damaged region is different in nature (rather than reflecting a change of kinetics) to that of the undamaged region.

Neglecting radiolysis effects which will be considered later, only α-and fission events will produce significant lattice disruption in waste forms. Some results calculated by the Kinchin-Pease method (9) by Roberts et al. (10), Malow and Andresen (11) and Antonini et al. (12) are given in Table 20. Calculations have been made of the evolution with time of the α- and β-productions for reprocessed waste (11) and the number of α-decays·$cm^{-3}$ in a reference reprocessed waste form (20 wt.% of six-year-old waste oxides in borosilicate glass) (11). Similar calculations have been made by Ringwood et al. (13) using the prescription of Cohen (14).

### TABLE 20

#### Atomic Displacments Due to Different Radiations

| Radiation | No. of Displacements | Maximum* Displacements | Principal Energy Transfer Mechanism | Refs. |
|---|---|---|---|---|
| α-particle (E = 6 MeV) | ∿100 | $2 \times 10^5$ | Ionization except at very end of trajectory | (10-12) |
| Associated α-recoil (E = 0.1 MeV) | ∿2000 | $4 \times 10^3$ | Displacement/ionization | (10-12) |
| >0.5 MeV β⁻ average = 1.5 MeV | <0.1 | $6 \times 10^4$ | Ionization | (10-12) |
| 2 MeV γ | <<0.1 | $8 \times 10^4$ | Ionization | (10) |
| 46·5 MeV Ni$^{6+}$ | $8 \times 10^3$ | $2 \times 10^6$ | Ionization | (12) |
| $^{10}B(n,\alpha)^7Li$ + 2.8 MeV | ∿800 | $1 \times 10^5$ | As for α-particle | (11) |
| $^{235}U(n,f)f'$ + 168 MeV | ≈$10^4$ | $7 \times 10^6$ | Ionization | (10-12) |
| Fast neutron | 200 | $4 \times 10^4$/MeV | Ionization | (12) |

*(Incorrectly) assuming (a) all energy used for bond-breaking and (b) displacement energy = 25 e.v.

To decide at the outset whether α-recoil damage effects might be significantly deleterious to a solidified nuclear waste form, probably the most direct information for crystalline ceramic formulations comes from studies of metamict minerals. This subject has been reviewed several times (15-18). Lattice damage in a mineral accrues from the decay of α-emitting U, Th and daughters incorporated in the lattice of the initially crystalline mineral. Probably the best known example is the mineral zircon ($ZrSiO_4$), which is rendered amorphous by a fluence (deduced from the U/Pb age and the radioactive content) of ∿$4 \times 10^{19} \alpha \cdot cm^{-3}$ (19), just the order of magnitude of the $10^4$-$10^6$ yr. dose to the reference reprocessing waste solid (11) mentioned previously. [That the metamict mineral can be recognized as such relies on the observation of crystalline morphology in cases where weathering has not been severe and/or the development of a recognizable x-ray pattern on annealing in a suitable atmosphere at ∿1000°C.] Thus if the α-emitting nuclei in a solid waste form were concentrated in a zircon-like phase, problems of considerable volume expansion [∿16% (19)], and more seriously, enhanced leachability (17,18) could perhaps occur.

In general, however, it will be unlikely that there will exist any direct evidence on radiation effects in natural analogues of the phases actually in a candidate waste form, so laboratory studies must be made. Whereas the natural process of metamictization takes millions of years (the dose rate will be within a factor of $\sim 10^2$ [slower] of the average rate in a candidate waste form), it is obviously desirable that a laboratory experiment takes place in a time of the order of 1 year or less. In this case the dose rate will be $\geq 10^5$ times the average rate in the candidate waste form. It is assumed that the radiation response of the waste form over a period of $\geq 10^5$ yr. needs to be modeled.

Further problems may of course arise in relating the laboratory data to results expected for the candidate waste form: unless the mechanisms of the radiation effects can be pinpointed, large uncertainties will remain in the projected behavior of the candidate waste form. However, studies of irradiation temperature effects, dose rate effects and chemical variability of the samples would go a long way towards minimizing these uncertainties.

In the study of $\alpha$-recoil effects, incorporation of an $\alpha$-emitter such as $^{238}$Pu or $^{244}$Cm ($t_{\frac{1}{2}}$ = 87 and 18 yr., respectively) in the solid of interest probably constitutes the most "direct" experiment. However, the handling of these (costly) actinides requires sophisticated facilities. Fortunately a whole range of other possible irradiation techniques exists, though there is somewhat of a trade-off between experimental convenience and relevance to the waste disposal problem. These irradiation techniques are: fast neutrons, fast neutron fission of $^{235}$U and $^{238}$U, slow-neutron fission of $^{235}$U, slow neutron reactions on $^{10}$B or $^{6}$Li, accelerated ions or 'external' $\alpha$-sources. The more superficial advantages and disadvantages of these methods are set out in Table 5, together with some relevant references. One major advantage of all these techniques relative to actinide doping is that in most cases the sample radioactivity is modest.

For the study of ionizing radiation alone, doses of $\sim 10^{12}$R (1) are needed to produce the long-term dose after solidification. Such experiments can be carried out on a micro-scale by the electron microscope method or on bulk samples with a Van de Graaff electron accelerator or $^{60}$Co source.

## Interaction of Massive Particles with Solids

### Basic Principles

In solids not subject to radiolysis (see below), $\beta$- and $\gamma$-rays essentially produce only ionization effects (electron-hole pairs), whereas relatively massive energetic particles like $\alpha$-particles, $\alpha$-recoil nuclei, fast neutrons and fission fragments can produce relatively large numbers of displaced atoms (Table 20). An important feature for a solid medium traversed by a moving massive particle is that if the particle is nominally an atom and is moving with a velocity greater than the orbital velocity of its outer electrons, these outer electrons will be stripped—the particle becomes ionized. The ion will, of course, interact with the electrons in the solid causing electron displacements but it will be repelled by the nuclei in the solid, so relatively few "billiard-ball" impacts on nuclei can occur. However, by virtue of the ion's interaction with the electrons in the solid it will eventually slow down to a point where it regains its lost electrons and becomes an atom again. At this stage the Coulomb barrier between it and the nucleus is essentially removed (even though the atom is not a "point" entity like a neutron) so that atomic displacements occur more frequently near the end of the trajectory. The damage cross-section along the trajectory is approximately described by the well-known Bragg curve. A recent discussion has been given by Fleischer et al. (39).

The atomic displacement calculations quoted in Table 20 ignore annealing effects at the irradiation temperature and ionization-induced displacements (23,39). If it is crudely assumed (a) that self-annealing is negligible and (b) that a material is rendered amorphous at a dose corresponding to 1 displaced atom per atom in the solid, the derived number of displaced atoms per $\alpha$-recoil event for zircon is $\sim 10^3$ (25,40), a figure in good agreement with the corresponding number in Table 20. However, for fission-fragment irradiation of zircon, zirconia, $U_3O_8$ and $\alpha$-alumina, which are all reasonably refractory solids in which assumption (a) might be reasonable, it has been deduced (23-26) that each fission event produces $10^6$-$10^7$ displaced atoms, a value much higher than that given in Table 20. Similar figures apply to monazite [$(Ce,La)PO_4$], huttonite ($ThSiO_4$), xenotime ($YPO_4$), zirconolite ($CaZrTi_2O_7$), and perovskite ($CaTiO_3$), since the fission doses required to render each of these materials amorphous are fairly close to that for zircon (40). The discrepancy in the number of displaced atoms per fission event between the experimental results (which must give minimum numbers if some annealing can actually occur) and the calculated results (Table 4) shows that ionization effects at these very high particle energies can cause atomic displacements. Again, these effects have been discussed qualitatively by Fleischer et al. (39,41).

Extensive studies of ion irradiation effects on materials in general have been made by Naguib and Kelly (31): various criteria for radiation-susceptibility were given. The important features for radiation susceptibility of *refractory* materials appear to be large open spaces in the structure, non-cubic crystal structures and high crystallization temperatures.

For the $\sim 4$ MeV $Kr^+/Ar^+$ irradiation experiments on zircon by Cartz and co-workers (32-34), a calculation similar to that above for zircon gives again $\sim 10^3$ displaced atoms per $Kr^+/Ar^+$ ion. This value is comparable to that for the $\alpha$-recoil case in this material, suggesting that, unlike the fission fragment technique (36,40) the $Kr^+/Ar^+$ ion experiments are probably a good means of simulating the $\alpha$-recoil phenomenon. Cartz et al. (32-34) have also studied structural effects on monazite, huttonite, $\alpha$-quartz, $LiAlSiO_4$, $ThO_2$ and various phosphates and nitrides. The $\sim 200$ KeV irradiation experiments of the French group (28-30) who use Pb ions should be an even better simulation in principle of the $\alpha$-recoil process, although difficulties will be encountered with the very short range ($\sim 500\text{\AA}$) of these particles.

There are virtually no data available for the slow-neutron reactions on $^6Li$ or $^{10}B$ but it would be expected that the displacement efficiencies would be approximately those of 5 MeV $\alpha$-particles because of the rough

similarities of the energies and masses of the relevant particles. It is
not clear whether the figure of Malow and Andresen (11)--see Table 20--
was actually derived from experiment or was calculated. For cubic BN,
unpublished work (21) showed that the amorphous state was reached after a
slow neutron fluence of $\sim 3 \times 10^{20}$ $n \cdot cm^{-2}$--this corresponds to approximately
10% atomic burnup. Assuming the foregoing crude relation between damage
and amorphism, each $^{10}B$ transmutation produces $\sim 10$ displaced atoms. This
number is lower, by an order of magnitude or so, than that expected for 5
MeV $\alpha$-particles (Table 4). Similar calculations for $B_4C$ give even lower
numbers of displaced atoms, as a slow neutron fluence of $\sim 3 \times 10^{20}$ did not
produce amorphism, although the lattice was heavily damaged (20). The
discrepancy is probably attributable to defect recombination (self-
annealing) effects occurring during irradiation. The transmutation process
itself (see later) no doubt also contributed to the lattice damage.

Many calculations have been made for fast neutrons. These almost all
rely essentially on the Kinchin-Pease approach (9). Detailed discussions
have been given some years ago [see Refs. (27) and (42), for example].
However, recourse still has to be made to experiment. Though neutron
irradiation is widely used, very few refractory materials have been
rendered amorphous by fast neutrons alone, in part for the practical reason
of neutron flux limitations. Diamond and quartz are well known and
probably the only examples of such materials. However, for quartz, there
is the possibility of complications with $OH^-/H_2O$ impurities (see below).
For diamond, the fast neutron fluence for x-ray amorphous material to be
produced is $\sim 10^{21}$ $n \cdot cm^{-2}$ (43,44). Making again the crude calculation, it
is easily found that the number of displacements per primary neutron is
$\sim 100$. In the experimental study, however, considerable variability from
diamond to diamond was observed (43). It is possible that boron impurities
(B is a well-known impurity in diamond) may have provided an additional
contribution to the damage via the slow neutrons accompanying the fast
neutrons.

The volume expansion vs. neutron fluence is, unlike the case for
natural zircon (in which the damage is due to $\alpha$-recoils), not even close to
linear (43,44). Keating (44) deduced from studies of diamonds irradiated
at relatively low doses, such that the diamonds were still largely crystalline, that approximately 100 atoms were displaced per primary neutron,
but that only 2.5 of these were permanently displaced. It is not clear
what models were used in these calculations.

Radiolysis

The radiolysis problem of interest here is the atomic disruption of
materials by purely ionizing radiation. Theory gives a good description
of defect production by ionizing radiation in light alkali halides.
Absorption of a photon by a halogen ion from the radiation field creates a
bound electron plus a neutral halogen; this halogen combines with another
halogen ion to form a molecular-ion in an excited state and the decay of
this entity to the ground state causes a movement of the nuclei. In
susceptible materials, the nuclear movement can cause displacements (45).

Following earlier work by Das and Mitchell (46), Hobbs and co-workers
(47-49) have rendered $\alpha$-quartz amorphous by electron irradiation; from the
energy dependence of the process they attribute the metamictization to

radiolysis, not classical atomic displacement. It is argued (48) that these metamictization effects in silicate-based waste forms might be as important as α-recoil effects. Quantitative justification of this proposal is presently lacking, and the results are apparently influenced by the $H_2O/OH^-$ content of the quartz. Radiolysis effects in irradiated zeolites and MgO due to $OH^-$ and $H_2O$ are well known (50,51), but the amount of water in a dry-fired glass or ceramic should be negligible. Although recent work on natural quartz shows that electron-beam metamictization nucleates at dislocations (52), the mechanical properties, which will be directly related to the dislocation structures, of naturally-stressed quartz are known to be dependent on $H_2O/OH^-$ content (53,54).

## Effects of Radiation in Solids

General discussions of radiation effects in solids are numerous, so only broad features are briefly outlined here. All crystalline solids contain lattice defects such as solid solution impurities, grain boundaries, interstitials, precipitate particles, surfaces, vacancies, etc. These defects perturb the otherwise perfectly periodic charge distribution and the defective lattice can be regarded as the sum of a periodic charge distribution plus a set of impurity charges.

### Insulators

When electron-hole pairs are produced by ionizing radiation, the displaced electrons or the holes can be trapped by the impurity charges. These systems constitute color centers though of course if the principal allowed electronic transitions from the ground states to the excited states do not lie near the optical energy band (1.8-3.0 e.v.) no actual coloration will be observed. When the solid, however, is heated or irradiated with photons of suitable energy the electrons and holes can be untrapped (ionization of the defects) and the color centers are said to be bleached. With purely ionizing radiation, and if no radiolysis can occur, the number of the color centers is limited by the defect content of the solid, not the intensity or total dose of the radiation.

The most detailed studies from an atomistic point of view have been made on defective simple solids which can be made with very high chemical purity (so that intrinsic defect effects can be distinguished from impurity effects). The solids in question are the light alkali and silver halides in which purely ionizing radiation can produce various kinds of atomic-scale defects. Most of the definitive experimental studies have been of a spectroscopic nature--electron spin resonance, especially of the optically-excited state, Zeeman and uniaxial-stress perturbation spectroscopy at low temperatures and various modulated versions of these. As a result, the properties of isolated single defects, and of small aggregates, are generally well characterized by their site symmetries and electronic energy levels.

In general however, in more "useful" materials, one of the most popular means of producing radiation damage is to expose it to fast neutrons in a nuclear reactor. Relatively massive samples can be studied (Table 5). Obviously the most detailed studies have been carried out on reactor fuels, structural materials and moderators. For a review of work in this direction up until about 1960 see Ref. (27).

TABLE 21

Various Kinds of Atom Displacement-Producing Techniques

| Type | Approximate Neutron cross-section (barns) | Natural Isotopic Abundance (%) | Range | Relevant References |
|---|---|---|---|---|
| (1) $^{10}B(n,\alpha)^7Li$ + 2.8 MeV | 4000 (slow) | 19.8 | ~ 2 μm | (20-22) |
| (2) $^6Li(n,\alpha)^3H$ + 4.8 MeV | 900 (slow) | 7.5 | ~ 5 μm | (22) |
| (3) U(n,f)f$^1$ + 170 MeV | 600 (slow) $^{235}U$ | 0.7 $(^{235}U)$ | ~10 μm | (23-26) |
| | 300 (fast) $^{235}U$ | | | |
| | ~$10^{-3}$ (fast) $^{238}U$ | 99.3 $(^{238}U)$ | | |
| (4) Fast neutrons (≥1 MeV) | ~3 | | ~10 cm | numerous-see e.g. (27) |
| (5) Accelerated ions $(H^+, D^+, T^+, He^+, Kr^+, ...)$ | -- | | ≳10 μm (highly dependent on mass and energy | numerous-see (28-34) |
| (6) 'External' α-sources $(^{210}Po, ^{244}Cm)$ | -- | | ~10 μm | (35,36) |
| (7) Spontaneous fission sources $(^{252}Cf)$ | -- | | ~10 μm | (37,38) |

Comments

(1) B may be difficult to dissolve in radwaste phases--unless dispersed at <1 μm, self-absorption of $^7Li$, α in a B-rich second phase is a problem. However, B may substitute for Al in some materials, especially if the Al is tetrahedrally coordinated.

(2) Same as (1), unless phase of interest contains alkali metals or, possibly, alkaline earth oxides.

(3) Even if insoluble, U can be readily dispersed on the ~2 μm scale in ceramic phases so that the self-absorption problem is relatively unimportant.

(4) Wide energy spectrum and different atomic masses and scattering cross sections in compounds makes detailed calculations difficult.

(5) Unless using very light, very energetic ions, only a thin layer of target is irradiated--so substrate epitaxy complications ensue. Damage profile somewhat non-uniform even in irradiated layer, leading to characterization difficulties.

(6) As for (5). There is an intrinsic dose limit which can be produced by a single source, because only α-emitting atoms within ~10 μm of the surface of the source can provide an α-particle which may emerge from the source.

(7) As for (6); fission flux ~3% of α-flux.

As far as the irradiated (crystalline) solids are concerned, the most studied properties are structural and dimensional changes. Dimensional changes can be very large, in excess of 10%, for heavily irradiated materials (19,43). A recent discussion of transmission electron microscopy observations on irradiated materials has been given by Hobbs (47). From spectroscopic studies, it can be deduced that at low doses, relatively isolated single atoms are displaced to form Frenkel defects, with the vacancy existing in a variety of charge states. It is the anion lattice rather than the cation lattice which is usually prone to suffer displacements, both because the anions are generally lighter than the cations and because the anions generally have a lower valence--thus the coulomb energy of an anionic Frenkel defect is less than or equal to that of a cationic Frenkel defect. The solid as a whole normally increases in size, because an interstitial causes more expansion than a vacancy causes contraction, other things being equal. The Bragg peaks in x-ray diffraction are somewhat weakened and diffuse scattering appears because of the deviations from periodicity arising from the presence of Frenkel defects. At higher doses and/or irradiation temperatures, the defects minimize the lattice strain by aggregation into multiple-atom defects and condensation into dislocation loops. At this stage, x-ray peaks become diffuse and cease to give a definite indication of the average lattice parameter (43,55-57). The x-ray crystallography of radiation-damaged structures has been dealt with in detail by Krivoglaz (56) and more recent discussion has been given by Dederichs (58); see also (59).

As implied earlier, a very large literature exists on metamict minerals such as zircon, allanite, etc., but these studies are by and large quite phenomenological compared to those on synthetic minerals because of the chemical complexity of most natural minerals.

## Metals

Generally speaking, the atoms in metals and alloys are more mobile than they are in refractory insulators so that self-annealing at room temperature is nearly complete. However, the mobile defects created in a radiation field (of massive particles) lead to enhanced creep in load-bearing alloys. Radiation also enhances corrosion (probably mainly via its effects on the fluid) and may change appreciably the kinetics of order-disorder and precipitation reactions in alloys. Since optical and micro-wave spectroscopies in the ordinary sense are not viable for the study of radiation defects in metals, basic studies have primarily relied on low-temperature ($\leq 4.2°K$) irradiation to minimize self-annealing and have mostly used defect electrical resistivity as an index of the damage; elaborate annealing studies over the 4-200K range have been reported for neutron-irradiated metals [see, e.g., (27)].

Other features which may play a role in nuclear waste forms are interstitial helium buildup from stopped $\alpha$-particles, and voids which may form in materials irradiated at somewhat elevated temperatures. Permar and McDonell (1) give a qualitative discussion, from which they conclude that He accumulation ($\leq 0.1$ atomic % in $10^6$ yr. for reprocessing waste) will be a minor problem. The observation by transmission electron microscopy of radiation-induced voids has been discussed in various materials by many authors. Void formation is favored by the moderately-elevated temperatures as would exist during the decay of $^{90}Sr$ and $^{137}Cs$ in a reprocessing waste form. Though void formation and swelling are phenomena that can be observed at radiation doses typical of those expected in reprocessing wastes after $10^6$ yr. (47,60), the effects will clearly be waste form-dependent.

## Discussion of Work on Candidate Waste Forms

Several compilations and discussions of published work relevant to radiation effects on candidate nuclear waste forms were produced very recently (1,61-65) so here the aim will be to discuss the work in a rather more discursive fashion than has been done previously.

Commercial Reprocessing Waste Forms

In prototype HLW borosilicate glasses (both vitrified and devitrified) which were self-irradiated via $^{244}$Cm incorporation, corresponding to α-doses occurring after ∿$10^5$ yr. (but see below), the density changes were observed to be ≤1%, and the leach rate did not change very much in Soxhlet testing or under mildly acid or alkaline conditions (66,67). Other studies in the U.K. on $^{238}$Pu-doped radwaste glasses support these results--here the leaching was done at room temperature and with distilled water or seawater (68) or with distilled water at 0-100°C (69). No substantial leaching changes were observed with other waste glasses irradiated to a γ-dose of $10^{11}$R (68).

No significant impairment of physical durability was observed in European work using 46.5 MeV Ni$^{6+}$ ion irradiation and fission fragment irradiation (12,70) or the $^{10}$B(n,α)$^7$Li reaction (11). However, in these studies, leach resistance was not among the physical properties examined. The dose rates were calculated from data similar to those shown in Table 4 to be equivalent to >$10^6$ yr. of post-solidification storage, but the fission fragment irradiation (as we have argued previously) and very probably the 46.5 MeV Ni$^{6+}$ irradiations would provide rather poor simulations of the α-recoil process.

The recent results of the French group (28-30) appear to be at variance with the above. The use of 200 KeV Pb ions should be a very good way to simulate α-recoil irradiation, as mentioned earlier. The initial results (28,29) on borosilicate waste glasses suggested that until an α-recoil dose corresponding to ∿2000 yr. of post-solidification storage was accumulated, little change in the leach rate would result, but beyond this dose, a catastrophic increase in the leach rate would occur. The leaching solution was generally 250 gm/litre NaCl solution at 100°C. However, the leach rate enhancement for a given glass after irradiation was sensitive to the NaCl concentration and pH of the leaching solution. The most recent results (30) suggest the picture to be somewhat more complicated. Firstly, the leach rate enhancements produced by irradiation were very variable. For one borosilicate glass formulation, no acceleration in the leach rate was observed even after high doses of Pb ions. But, overall, the data, both on minerals and waste glasses, suggest that the catastrophic leach rate increase is linked to the corrosive nature of the leachant; that is, the more corrosive the leachant, the more likely will a catastrophic leach rate change occur. This result is not of course unexpected, in the light of remarks previously made in connection with the specific and very corrosive natures of etchants used to develop fission tracks in irradiated minerals (see the first page of this chapter). Note that glasses per se are in no way immune to fission damage (71).

Of the large number of glasses studied (30), only one was a borosilicate glass containing simulated radwaste. Perhaps coincidentally, this glass did not show a catastrophic leach rate increase after the standard irradiation treatment.

A number of possibilities may explain why only the French group has observed these catastrophic leach rate enhancements. They themselves point out obvious shortcomings of the technique, i.e., the confinement of the damaged region to the surface (∿500Å thick layer) and the fact that the leach rate itself applies only to this thin surface layer, so that it may not be relevant to long-term leaching behavior. Other possibilities are a potential high sensitivity to glass chemistry and problems with perturbing the glass surface by polishing.

One problem for the actinide-doped glasses which has not explicitly been commented on is the possibility of lowered $\alpha$-recoil doses to the glass by non-uniform doping. The simulations must contain more actinide than the "real" glass to make long-term predictions; non-uniform doping or phase separation will therefore be more likely in the simulation experiments. In support of this position, Bonniaud et al. (72) have observed $\sim$1 $\mu$m-sized actinide-rich regions in Cm and Pu-doped glass formulations. However, Weed et al. (73) observed no segregation in PNL waste glass simulations by $\alpha$-autoradiography, and Walker and Riege (74) found solubilities of several percent of actinides in their formulations (the presence of Gd, however, considerably lowered the solubility of Pu).

$\alpha$-radiography is actually an insufficient criterion of actinide distribution as this distribution can in this way only be explored on a $\sim$1 $\mu$m scale. Since the range of an $\alpha$-recoil is only about $\sim$100Å, for "self-absorption" of actinide recoils not to occur it would be necessary to establish that the actinide ions are distributed on a scale of $\leq$100Å. Measurements using $\alpha$-recoil track development (39,75) are necessary to examine this point. For individual *crystalline* phases the solubility of some of the actinides on an atomic scale can be established by spectroscopic methods (25,26,76), but for a nuclear waste glass which would contain many paramagnetic impurities this is probably not feasible, though spectroscopic studies of U-containing glasses have been made recently in the radwaste context (77).

Hirsch (78) has reported severe degradation of the leach resistance of borosilicate glasses using doses of $\sim$10$^{16} \cdot$cm$^{-2}$ of 1.6 KeV Ar$^+$ ions. Here the penetration will be only a few tens of Å but the energy deposition in this layer would correspond to $\sim$10$^2$ times that component characterized by displacement damage sustained after $\sim$10$^6$ yr. in the reference reprocessing form. Thus this experiment is not very relevant.

Radiation data on non-vitreous candidate forms are very sparse indeed. This is largely because the development of these forms is still being pursued--the technology is "younger" than that of glass--radiation resistance is just one design parameter, as mentioned earlier. The PSU tailored-ceramic (79-81) doped with $^{244}$Cm was investigated for phase constitution and density (66,81) as a function of self-irradiation time but no leach data have been reported. A devitrified PNL waste glass simulation however, was shown to exhibit only a minimal leach rate enhancement after $\sim$10$^3$ yr. of equivalent post-solidification storage (66). A neutron-irradiated Synroc B preparation (82,83) showed no microcracking, while single phase hollandite and perovskite preparations did (84); the irradiation was nominally equivalent to a dose of $\sim$10$^4$ yr. post-solidification--some density results were recorded. Most work on radiation damage of tailored ceramics has been done on individual phases, and not by the actinide doping technique. In the laboratory, phases occurring in the PSU materials and Synroc (82,83), such as monazite, perovskite, zirconolite, uraninite, thoria, etc., have been studied by fission fragment (23,40) and $\sim$4 MeV Ar$^+$/Kr$^+$ ion irradiation (32-34), in respect of structural changes.

However, enhanced reactivity with aqueous media at 200°C in deionized water for some fission-fragment irradiated materials was deduced (40), though more accurate $\alpha$-recoil simulations are necessary.

At this stage, most of the work on the radiation effects on solubility properties of these phases relies on geological evidence. For monazite, geological evidence strongly points to its leach resistance even when it contains substantial amounts of U and Th (1,17,18,85). Even the presence of Ca, which might be expected to be readily leachable, does not drastically remove its leach resistance--no changes in the x-ray pattern of a powdered ceramic preparation of monazite-structured $Ca_{0.5}U_{0.5}PO_4$ were observed after a treatment at 200°C/30 MPa for 21 days in deionized water (86). Uraninites of various geological ages seem to be very leach resistant in reducing ground waters but they may be susceptible under oxidizing conditions (87-92). However, $U^{6+}$ or even $(UO_2)^{2+}$ is not necessarily very soluble; carnotite, in which $(UO_2)^{2+}$ is a major constituent, is actually very insoluble (92). Experiments in this laboratory are in progress to study leaching effects in uraninite-based single-phase fluorites in which the oxygen/metal content and thus the inferred $U^{6+}/U^{4+}$ ratio, is varied. Ringwood et al. (14,93) have made detailed isotopic and chemical analyses of natural zirconolites and have argued that even x-ray amorphous samples have not suffered leaching. Moreover, they have performed leaching experiments at 200°C on naturally-irradiated zirconolites and find the leach rate, based on U analysis of the leachate, to be quite low. In agreement with this, amorphous synthetic zirconolite showed no alteration after 7 days at 200°C/2 MPa in deionized water (40); under these conditions other synthetic minerals which were rendered amorphous did show alteration.

## High-level Defense Waste

Since defense wastes have low radioactive loadings, $\geq$100 times lower than those of commercial reprocessing wastes, it is generally agreed that radiation effects in non-partitioned defense wastes will be minimal, especially if borosilicate glasses are employed--the present reference process. However, experiments on candidate-formulations will still be necessary. Few such experiments have been carried out.

A simulated Savannah River waste sludge diluted 1:3 by weight with borosilicate glass and in which the $\alpha$-emission was produced by incorporation of $^{238}Pu$ or $^{244}Cm$ (94) showed only very small density changes after an equivalent dose of $\geq$10$^6$yr. of post-solidification storage. No significant changes in leach rate were observed, but some microcracking took place.

Some work has been reported on simulated Savannah River waste incorporated in cement (95,96). Here one possible radiation problem is radiolytic gas generation, principally hydrogen and oxygen. Data in this respect have also been obtained for FUETAP concretes (97). Work has also been done on a first-generation tailored ceramic designed for Savannah River waste (40), but this fission-fragment irradiation study was, again, a poor simulation of the damage expected in an actual waste form.

# TRANSMUTATION EFFECTS

## Basic Principles

Table 17 shows the fission products derived from U-based reactor fuel and Table 22 shows the key elements for which transmutation problems will arise when the nuclides are solidified. The transmutation problem in a solid waste form derives from the fact that when a nuclide undergoes radioactive decay, the newly-formed atom may be a chemically different species, having a different ionic size and, possibly, valence. Thus if the nuclide is chemically incorporated in a solid form before transmutation it may not fit into the lattice after transmutation has occurred.

TABLE 22

Relatively Abundant Fission Product Elements for Which a Valence Change Will Occur in a Solid Waste Form and Which Have $t_{1/2} < 10^6$ yr

| element | stable daughter | $t_{1/2}$ (yr) |
|---------|-----------------|----------------|
| Cs | Ba | 30 |
| Sr | Zr | 29 |
| Kr* | Rb | 11 |

*Will not exist in general waste stream.

In typical nuclear reprocessing wastes the most significant transmutations are

$$^{137}_{55}Cs^+ \xrightarrow{\beta} {}^{137}_{56}Ba^{2+} \ (t_{\frac{1}{2}} = 30 \text{ yr})$$

and

$$^{90}Sr^{2+} \xrightarrow{\beta^-} {}^{90}Y^{3+} \ (t_{\frac{1}{2}} = 29 \text{ yr}) \xrightarrow{\beta^-} {}^{90}Zr^{4+} \ (t_{\frac{1}{2}} = 64 \text{ hr}).$$

For the latter decay series, the amount of $Y^{3+}$ relative to $Zr^{4+}$ is negligible.

In both the $Cs^+ \rightarrow Ba^{2+}$ and $Sr^{2+} \rightarrow Zr^{4+}$ transmutations the ionic size differences are 20-40%. This alone poses problems, but there are several structures in which the ionic radius tolerance factor is substantial--for example the XII-fold site in perovskite has been known for many years to tolerate a 40% size change.

A mechanism for charge compensation is another matter and both crystalline and non-crystalline phases (insulators) share the same problem, which must be resolved by corresponding changes in other cations or anions. [No transmutation problem due to valence changes occurs in a metal--"electrical neutrality" does not apply. However, a metal containing appreciable Cs or Sr would be very reactive with water.] For insulating crystals it seems clear that some proportion ($\sim$1%) of transmutation could be accommodated but that disproportionation or "decomposition" would take place after a certain proportion of the ions had become transmuted. If this process took place at a sufficiently low temperature (preventing appreciable atomic diffusion) the lattice strains around the transmuted (now misfitting) ions would cause smearing-out of the x-ray diffraction pattern much in the same way as does radiation damage, cold-work or glass formation. The transmutation-induced departure from thermodynamic equilibrium would increase as transmutation continued, and it would be a question of kinetics when phase decomposition would occur.

Two lines of experimentation are possible to study the possible effects of transmutation on solid waste forms. These are: (a) fundamental transmutation studies using radioactive or neutron-induced transmutation; (b) chemical studies of phase stabilities when transmutation is simulated ("before and after" investigations), with emphasis for the problem of nuclear waste disposal on *mitigation* of phase disproportionation or decomposition.

## Fundamental Transmutation Studies

Virtually no relevant studies have actually been performed, though one (35) is in progress. To study transmutation effects on the structures of crystals, diffraction studies are vital. Although isolated (transmuted) cation studies in solids can be made with techniques such as $\gamma$ angular correlation, Mössbauer, optical, and paramagnetic resonance spectroscopies, this information does not directly relate to the 'collective' state of the solid and its phase constitution. Ten percent of transmutation of a major element in the phase of interest should be sought as a minimum in fundamental studies. In such studies, both radioactive and neutron-induced transmutation would seem worth pursuing. The advantage of (reactor) neutron-induced transmutation is that the desired sample can be prepared using stable isotopes so that there are no constraints on sample preparation, either in terms of material volume or of preparation time. The disadvantages are the long irradiation times required to produce an adequate degree of transmutation and possible radiation damage complications from fast neutron fluxes.

Other considerations are as follows:

(a) parent and daughter must not be variable-valence transition metals or actinides (see below);

(b) after neutron capture, transmutation should preferably take place in a reasonable time and via $\beta^-$, $\beta^+$ or electron-capture, not by fission or evolution of massive particles which would introduce additional radiation damage;

(c) no other nuclei, including the daughter of the compound nucleus, should undergo any significant transmutation.

Even with quite a high thermal neutron flux, say $10^{14}$ n cm$^{-2}$·sec$^{-1}$, for 10% transmutation the product of the isotope abundance and the particular absorption cross-section must exceed 30 barns, if irradiation times are not to exceed one year.

Favorable cases appear to be

$$^{133}_{55}\text{Cs}^+ + {}_0\text{n}^1 \rightarrow {}^{134}_{55}\text{Cs}^+ \xrightarrow{\beta^-} {}^{134}_{56}\text{Ba}^{2+}$$

and

$$^{174}_{72}\text{Hf}^{4+} + {}_0\text{n}^1 \rightarrow {}^{175}_{72}\text{Hf}^{4+} \xrightarrow{\varepsilon} {}^{175}_{71}\text{Lu}^{3+} \ .$$

However, the induced radioactivities would be formidable for $\sim$10% transmutation, the specific activities of possible host compounds being on the order of 100 Ci/gm on withdrawal from the reactor and the principal half-lives being 2.0 and 0.2 years, respectively, for the two cases above.

These levels of activity will surely tax x-ray diffraction apparatus, even with large amounts of shielding to reduce the radioactive background. *Electron* diffraction would be a preferred technique, especially as the *absolute* activity of an electron diffraction sample would be quite low (for general discussions of problems of diffraction on radioactive samples, see Refs. (98) and (99)).

## Chemical Mitigation Studies

Here the approach is to select multiple-cation solids for transmutation resistance, the idea being to select structures which can accommodate parent and daughter in the same site and which contain cations which can change valence to compensate for the transmutation-induced valence change.

For example, with the $\text{Cs}^+ \rightarrow \text{Ba}^{2+}$ transmutation we would seek a mineral phase solid solution of the type

$$\text{Cs}^+\text{R}^{z+}\text{O}_{(z+1)/2} \rightarrow \text{Ba}^{2+}\text{R}^{(z-1)+}\text{O}_{(z+1)/2} \quad (\text{R = transition metal ion}).$$

This approach to mitigation of the transmutation problem has been explored in some detail (100). Though simple in concept, it has not been found very easy to apply the strategy--so far, perovskite looks promising to deal with the Cs → Ba transmutation, but no system for the $\text{Sr}^{2+} \rightarrow \text{Zr}^{4+}$ transmutation has been found. Other possible solid phase hosts have also been discussed and it was also pointed out that the transmutation problem will not disappear in an amorphous waste form (100).

### The Transmutation Problem in Specific Radwaste Formulations

Dilution is an obvious means of easing the transmutation problem. Defense wastes are already very dilute in radionuclides so no important problems of phase instabilities due to transmutation are anticipated. However, formulations for reprocessing waste could have difficulties. The Ringwood (82,83) formulation is designed to be a dilute solution of radwaste in three major phases, Ba-hollandite, zirconolite and perovskite, so further dilution of the radwaste by the Ba, Ca, Zr, Al and Ti oxides comprising the 'solvent' could be achieved. Within limits, also, the mixture could be diluted by just the problem elements, *viz.* 'cold' isotopes of Cs and Sr.

On the other hand, the McCarthy (3,79-81) formulations were found in the present work not to be amenable to dilution by the tailoring additives, Ca, Sr, Al and Si oxides, insofar as additions of any combinations of excess $Al_2O_3$ and $SiO_2$ (added as $SiO_2$ sol and aluminum nitrate solution to the starting mixture) upset the phase assemblage. Nor could the formulation tolerate dilution with pollucite (again added in the form of stoichiometric mixtures of Cs, Al and Si oxides via $SiO_2$ sol and metal nitrate solutions). In both cases, the pollucite and scheelite phases (for Cs and Sr,Mo, respectively) were "lost," as far as x-ray diffraction was concerned. For $Al_2O_3$ and $SiO_2$ additives, some extra x-ray reflections were observed but for the "pollucite" addition, this was not the case. No great increase in Cs volatility occurred on firing, as deduced both from gravimetric work and from the observations that, on firing at 1100 or 1200°C, the x-ray diffraction patterns of air-fired materials were very similar to those of materials fired in welded platinum capsules.

## Summarizing Remarks

Radiation-effect/transmutation testing of candidate waste forms is still virtually unexplored, except, as we have seen in the foregoing, for borosilicate waste glasses. Even then, there is significant disagreement between new results and earlier results which relied on actinide doping. There has been some α-recoil testing (67) of tailored ceramic preparations with respect to phase structure plus further work on single phases (65). No published laboratory α-recoil data exists on Synroc (82,83), porous glass matrix (101,102) materials or any other candidate HLW forms except cements.

Various documents (63-65) have appeared recently on the variables in different formulations which may affect standardized radiation tests: it is not the object of the present exercise to detail this work. However, while actinide incorporation appears to be the currently favored radiation test (65), the opportunity is here taken to reiterate that admittedly crude calculations suggest that 4 MeV $Kr^+/Ar^+$ ion irradiation produces similar damage to the α-recoil. Given this result, the ease of the ion irradiation experiment versus the α-recoil experiment would be a strong point in favor of increasing the role of ion irradiation studies in waste form testing, from the radiation point of view.

It is clear that α-recoil effects will be an important contributor to radiation effects in candidate forms for reprocessing wastes. β/γ radiolysis effects *may* be important--experiments are necessary. Given that these radiolysis effects will probably be minimal in a refractory waste form which is anhydrous, it might be argued that if water contacts the waste form then a 'positive feedback' situation will arise. However, if the canister is able to shield the waste form from water for 1000 years (the basis of the NRC requirement) there will be no problem because virtually all the fission product activity will have decayed away at this stage.

For tailored crystalline ceramics, once single phases (and various chemical variations thereof) have been demonstrated to be suitably radiation resistant, then the design problem is to ensure that all the actinides enter these phases (1,103).

Metal containers will not be susceptible to $\gamma$-irradiation effects and if the canister is suitably thick, the radiation field at the outside of the canister should be minimal. Work is in progress at Sandia Laboratories on possible radiation effects on leaching of canister materials by aqueous media (104).

All the foregoing has ignored the questions of long-lived volatile elements ($^3$H, $^{14}$C, $^{85}$Kr, $^{99}$Tc, $^{106}$Ru, $^{129}$I, together with radiogenic and non-radiogenic alkalies and Mo). Depending on how the wastes are finally processed, these may or may not be recycled. Certainly $^{85}$Kr would be dealt with separately. Radiation effects in an anhydrous waste form should be inconsequential because $^{85}$Kr is a $\beta$-emitter. However, there is no way to balance the charge in the suggested solid immobilizing agents (105-107), and when $^{85}$Kr transforms to $^{85}$Rb, there will be the problems outlined earlier. Moreover, the production of alkali ($^{85}$Rb) will of itself render the form more liable to leaching, whether or not the Rb forms as the metal or as the positive ion. $^{129}$I, another $\beta$-emitter, would be another strong candidate for separate processing but the half-life is so long that even after $10^6$ yr., only 4% of the $^{129}$I would transform to $^{129}$Xe. It has been pointed out elsewhere that $^{129}$I presents no great hazard because of its low specific activity (108-110).

Spent fuel has been ignored too since no chemical tailoring to the waste itself is envisaged if the material is to be disposed of as such: the fuel would be highly leachable whether or not radiation effects were significant. The canister will be the first line of defense, working outward from the fuel itself.

Acknowledgements

Helpful discussions with R. Roy and K.K.S. Pillay are acknowledged. Of particular value were discussions held at the Materials Characterization Center Workshops on radiation effects in nuclear waste forms (Battelle Seattle Research Center, Seattle, Washington, July 29-30, 1980; August 13-14, 1981).

# REFERENCES FOR CHAPTER 8

1. P.H. Permar and W.R. McDonell (1980), Significance of Radiation Effects in Solid Radioactive Waste, DP-MS-80-27, E.I. duPont deNemours and Co., Savannah River Laboratory, Aiken, SC.

2. C.M. Lederer, J.M. Hollander, I. Pearlman, and V.S. Shirley (1979), Table of Isotopes, 7th Edition, John Wiley, NY.

3. G.J. McCarthy and M.T. Davidson, Ceramic Nuclear Waste Forms: I, Crystal Chemistry and Phase Formation, Bull. Amer. Ceram. Soc. 54, 782 (1975).

4. J.A. Stone, S.T. Goforth, Jr., and P.K. Smith (1979), Preliminary Evaluation of Alternative Forms for Immobilization of Savannah River Plant High-Level Waste, DP-1545, E.I. duPont deNemours and Co., Savannah River Laboratory, Aiken, SC.

5. G.J. McCarthy, Crystalline Ceramics from Defense High-Level Wastes, Nucl. Technol. 44, 451 (1979).

6. B.A. Staples, S. Pomiak and E.L. Wade (1979), Characteristics of Stored High-Level ICPP Waste Calcines, Ceramics in Nuclear Waste Management, Ed. T.D. Chikalla and J.E. Mendel, CONF-790420, Department of Energy Publication, p. 365.

7. K.M. Lamb, S.J. Priebe, H.S. Cole and B.D. Taki (1979), A Pelleted Waste Form for High Level ICPP Waste, Ceramics in Nuclear Waste Management, Ed. T.D. Chikalla and J.E. Mendel, CONF-790420, Department of Energy Publication, p. 224.

8. Radioactive Wastes at the Hanford Reservation, A Technical Review (1978), National Academy of Sciences, Washington, DC.

9. G.H. Kinchin and R.S. Pease, The Displacement of Atoms in Solids by Radiation, Repts. Prog. Phys. 18, 1 (1955).

10. F.P. Roberts, G.H. Jenks, and C.D. Bopp (1976), Radiation Effects in Solidified High-Level Wastes--Part I, Stored Energy, USERDA Report BNWL-1944, Battelle Pacific Northwest Laboratories, Richland, WA.

11. G. Malow and H. Andresen (1979), Helium Formation from $\alpha$-decay and its Significance for Radioactive Waste Glasses, Scientific Basis for Nuclear Waste Management, Vol. 1, Ed. G.J. McCarthy, Plenum Press, NY, p. 109.

12. M. Antonini, F. Lanza, and A. Manara (1979), Simulations of Radiation Damage in Glasses, Ceramics in Nuclear Waste Management, Ed. T.D. Chikalla and J.E. Mendel, CONF-790420, Department of Energy Publication, p. 289.

13. B.L. Cohen, High-Level Radioactive Waste From Light-Water Reactors, Rev. Mod. Phys. 49, 1 (1977).

14. A.E. Ringwood, V. Oversby, and W. Sinclair (1980), Effects of Radiation Damage in Long-Term Stability of Synroc Minerals, Scientific Basis for Nuclear Waste Management, Vol. 2, Ed. C.J.M. Northrup, Plenum Press, NY, p. 273.

15. A. Pabst, The Metamict State, Amer. Mineral. <u>37</u>, 137 (1952).
16. R.S. Mitchell, Metamict Minerals: A Review, Min. Record <u>4</u>, 177, 214 (1973).
17. R.C. Ewing and R.F. Haaker (1978), The Metamict State Radiation Damage in Crystalline Phases, <u>High-Level Radioactive Solid Waste Forms</u>, NUREG/CP-0005, p. 651.
18. R.F. Haaker and R.C. Ewing (1979), The Metamict State: Radiation Damage in Crystalline Materials, <u>Ceramics in Nuclear Waste Management</u>, Eds. T.D. Chikalla and J.E. Mendel, CONF-790420, Department of Energy publication, p. 305.
19. H.D. Holland and D. Gottfried, The Effect of Nuclear Radiation on the Structure of Zircon, Acta Cryst. <u>8</u>, 291 (1955).
20. C.W. Tucker and P. Senio, X-ray Scattering by Neutron-irradiated Single Crystals of Boron Carbide, I, Acta Cryst. <u>8</u>, 371 (1955).
21. E.R. Vance and H.J. Milledge, unpublished work.
22. M.C. Wittels and F.A. Sherrill (1959), Some Irradiation Effects in Non-metallic Crystals, <u>Advances in X-ray Analysis</u>, Vol. 3, Ed. M.W. Mueller, Plenum, NY, p. 269.
23. R.M. Berman, M.L. Bleiberg, and W. Yeniscavish, Fission Fragment Damage to Crystal Structures, J. Nucl. Mat. <u>2</u>, 129 (1960).
24. M.C. Wittels, J.O. Steigler, and F.A. Sherrill, Radiation Effects in Uranium-doped Zirconia, Reactor Science and Technology, J. Nucl. Energy, Pts. A/B <u>16</u>, 237 (1962).
25. E.R. Vance and J.N. Boland, Fission Fragment Damage in Zircon, Radiation Effects <u>26</u>, 135 (1975).
26. E.R. Vance and J.N. Boland, Fission Fragment Irradiation of Single Crystal Monoclinic $ZrO_2$, Radiation Effects <u>37</u>, 237 (1978).
27. D.S. Billington and J.H. Crawford, Jr. (1961), <u>Radiation Damage in Solids</u>, Princeton U.P., p. 233.
28. J.C. Dran, M. Maurette and J.C. Petit, Radioactive Waste Storage Materials: Their Alpha-Recoil Aging, Science <u>209</u>, 1518 (1980).
29. J.C. Dran, Y. Langevin, M. Maurette and J.C. Petit (1980), A Microscopic Approach for the Simulation of Radioactive Waste Storage in Glass, <u>Scientific Basis for Nuclear Waste Management</u>, Vol. 2, Ed. C.J.M. Northrup, Plenum, NY, p. 30.
30. J.C. Dran, M. Maurette, J.C. Petit and B. Vassent (1981), Radiation Damage Effects on the Leach Resistance of Glasses and Minerals: Implications for Radioactive Waste Storage, <u>Scientific Basis for Nuclear Waste Management</u>, Vol. 3, Ed. J.G. Moore, Plenum, NY, (in press).
31. H.M. Naguib and R. Kelly, Criteria for Bombardment-induced Structural Changes in Non-Metallic Solids, Radiation Effects <u>25</u>, 1 (1975).
32. L. Cartz, F.G. Karioris and R.A. Fournelle, Heavy Ion Bombardment of Silicates and Nitrides, Radiation Effects <u>54</u>, 57 (1981).
33. F.G. Karioris, K. Appaji Gowda and L. Cartz, Heavy Ion Bombardment of Monoclinic $ThSiO_4$, $ThO_2$ and Monazite, Radiation Effects Letters <u>58</u>, 1 (1981).
34. L. Cartz, F.G. Karioris, R.A. Fournelle, K. Appaji Gowda, K. Ramasami, G. Sarkar and M. Billy (1981), Metamictization by Heavy Ion Bombardment of α-quartz, Zircon, Monazite and Other Structures, <u>Scientific Basis for Nuclear Waste Management</u>, Vol. 3, Ed. J.G. Moore, Plenum, NY (in press).

35. W.J. Weber, J.W. Wald, and W.J. Gray (1980), Radiation Effects in Crystalline High-Level Nuclear Waste Solids, PNL-SA-8732, Battelle, Pacific Northwest Laboratories, Richland, WA.

36. W.J. Weber (1980), Ingrowth of Lattice Defects in Alpha-Irradiated $UO_2$ Single Crystals, PNL-SA-8574, Battelle, Pacific Northwest Laboratories, Richland, WA.

37. T.H. Gould and W.R. McDonell (1976), Radiation damage by $^{252}$Cf Fission Fragments and Alpha Particles, Radiation Effects and Tritium Technology for Fusion Reactors, Vol. II, USERDA Report CONF-750989, Oak Ridge National Laboratory, Oak Ridge, TN, p. 387.

38. W.R. McDonell and S. Dillich (1978), Effects of Radiation from Transuranium Nuclides on Container Surfaces, DP-MS-78-9, E.I. duPont de Nemours Co., Savannah River Laboratory, Aiken, SC.

39. R.L. Fleischer, P.B. Price and R.M. Walker (1975), Nuclear Tracks in Solids, University of California Press, Berkeley.

40. E.R. Vance and K.K.S. Pillay (1981), Experimental Study of Structural Damage from Fission Fragment Irradiation in Crystalline Nuclear Waste Phases, Scientific Basis for Nuclear Waste Management, Vol. 3, Ed. J.G. Moore, Plenum Press, NY (in press).

41. R.L. Fleischer, P.B. Price and R.M. Walker, Ion Explosion Spike Mechanism for Formation of Charged-Particle Tracks in Solids, J. Appl. Phys. 36, 3645 (1963).

42. G. Liebfried (1962), Radiation Damage Theory, Proceedings of the International School of Physics, Enrico Fermi Course XVIII, Ed. D.S. Billington, Academic Press, NY, p. 227.

43. E.R. Vance, X-ray Study of Neutron-irradiated Diamond, J. Phys. C: Solid St. Phys. 4, 257 (1971).

44. D.T. Keating, The Initial Phase of Damage in Neutron-Irradiated Diamond, Acta Cryst. 16, A113 (1963).

45. D. Pooley, F-centre Production in Alkali Halides by Electron-hole Recombination and a Subsequent [110] Replacement Sequence: A Discussion of Electron-hole Recombination, Proc. Phys. Soc. 87, 245 (1966).

46. G. Das and T.E. Mitchell, Electron Irradiation Damage in Quartz, Radiation Effects 23, 49 (1974).

47. L.W. Hobbs, Application of Transmission Electron Microscopy to Radiation Damage in Ceramics, J. Amer. Ceram. Soc. 62, 267 (1979).

48. M.R. Pascucci and L.W. Hobbs, Metamict Transformation in Silicate-Based Waste-Storage Media, Bull. Amer. Ceram. Soc. 59, 395 (1980).

49. L.W. Hobbs and M.R. Pascucci, Radiolysis and Defect Structure in Electron-Irradiated α-Quartz, J. Phys. Paris C6, 237 (1980).

50. Y. Chen, M.M. Abraham, L.C. Templeton and W.P. Unruh, Role of Hydrogen and Deuterium in the $V^-$-Center Formation in MgO, Phys. Rev. B11, 881 (1974).

51. L.A. Bursill, E.A. Lodge and J.M. Thomas, Zeolite Structures as Revealed by High-Resolution Electron Microscopy, Nature 286, 111 (1980).

52. D. Cherns, J.C. Hutchison, M.L. Jenkins, P.B. Hirsch and S. White, Electron-irradiation-induced Vitrification at Dislocations in Quartz, Nature 287, 314 (1980).

53. D.T. Griggs and J.D. Blacic, Quartz: Anomalous Weakness of Synthetic Crystals, Science 147, 292 (1965).

54. D.T. Griggs, A Model of Hydrolytic Weakening in Quartz, J. Geophys. Res. 79, 1653 (1974).

55. B.S. Hickman and D.G. Walker, Growth of Magnesium Oxide During Neutron Irradiation, Phil. Mag. 11, 1101 (1965).

56. M.A. Krivoglaz (1969), Theory of X-ray and Thermal-Neutron Scattering by Real Crystals, Plenum Press, NY, p. 249.

57. C.J. Howard and T.M. Sabine, X-ray Diffraction Profiles for Neutron-irradiated Magnesium Oxide, J. Phys. C: Solid St. Phys. 7, 3453 (1974).

58. P.H. Dederichs, The Theory of Diffuse X-ray Scattering and its Application to the Study of Point Defects and Their Clusters, J. Phys. F: Metal. Phys. 3, 471 (1973).

59. B.C. Larson and W. Schmatz, Huang Diffuse Scattering from Dislocation Loops and Cobalt Precipitates in Copper, Phys. Rev. B10, 2307 (1974).

60. F.W. Clinard, Jr. and G.F. Hurley (1979), Effects of Irradiation on Structural Properties of Crystalline Ceramics, Ceramics in Nuclear Waste Management, Ed. T.D. Chikalla and J.E. Mendel, CONF-790420, Department of Energy publication, p. 300.

61. D.W. Kneff (1980), Radiation Effects Simulation in High Level Nuclear Waste Forms: A Literature Review, Supporting Document No. 364T100001, Energy Systems Group, Rockwell International, Canoga Park, CA.

62. H.M. Lee (1980), Natural Radiation Damage in Solids: A Literature Survey Review, Supportive Document No. 364T1000002, Energy Systems Group, Rockwell International, Canoga Park, CA.

63. K.S. Czyesinski, K.W. Swyler and C.J. Klamut (1980), NRC Nuclear Waste Management Technical Support in the Development of Nuclear Waste Form Criteria. Task 4: Test Development Review, NUREG-77773 (informal report), Brookhaven National Laboratory, Upton, NY.

64. R. Dayal, K.W. Swyler and P. Soo (1980), NRC Nuclear Waste Management Technical Support in the Development of Nuclear Waste Form Criteria. Task 1: Waste Package Overview, Brookhaven National Laboratory, Upton, NY.

65. F.P. Roberts, R.P. Turcotte, and W.J. Weber (1981), Materials Characterization Center Workshop on the Irradiation Effects in Nuclear Waste Forms, Summary Report, PNL-3588, Battelle, Pacific Northwest Laboratories, Richland, WA.

66. W.J. Weber, R.P. Turcotte, C.R. Bunnell, F.P. Roberts, and J.H. Westsik, Jr. (1979), Radiation Effects in Vitreous and Devitrified Simulated Waste Glass, Ceramics in Nuclear Waste Management, Ed. T.D. Chikalla and J.E. Mendel, CONF-790420, Department of Energy publication, p. 294.

67. W.A. Ross, R.P. Turcotte, J.E. Mendel and J.E. Rusin (1979), A Comparison of Glass and Crystalline Waste Materials, Ceramics in Nuclear Waste Management, Ed. T.D. Chikalla and J.E. Mendel, CONF-790420, Department of Energy publication, p. 52.

68. K.D.B. Johnson (1979), The U.K. Program--Glasses and Ceramics for Immobilization of Radioactive Wastes for Disposal, Ceramics in Nuclear Waste Management, Ed. T.D. Chikalla and J.E. Mendel, CONF-790420, Department of Energy publication, p. 17.

69. K.A. Boult, J.T. Dalton, A.R. Hall, A. Hough and J.A.C. Marples (1979), The Leaching of Radioactive Waste Storage Glasses, Ceramics in Nuclear Waste Management, Ed. T.D. Chikalla and J.E. Mendel, CONF-790420, Department of Energy publication, p. 248.

70. M. Antonini, P. Camagni, F. Lanza and A. Manara (1980), Atomic Displacements and Radiation Damage in Glasses Incorporating HLW, Scientific Basis for Nuclear Waste Management, Vol. 2, Ed. C.J.M. Northrup, Plenum Press, NY, p. 127.

71. R.L. Fleischer and P.B. Price, Charged-particle Tracks in Glass, J. Appl. Phys. 34, 2903 (1963).

72. R.A. Bonniaud, N.R. Jacquet-Francillon and C.G. Sombret (1980), The Behavior of Actinides in α-doped Glasses as Regard to the Long Time Disposal of High-level Radioactive Materials, Scientific Basis for Nuclear Waste Management, Vol. 2, Ed. C.J.M. Northrup, Plenum Press, NY, p. 117.

73. H.C. Weed, D.G. Coles, D.J. Bradley, R.W. Mensing and J.S. Schweiger (1979), Leaching Characteristics of Actinides from Simulated Reactor Waste Glass, Scientific Basis for Nuclear Waste Management, Vol. 1, Ed. G.J. McCarthy, Plenum Press, NY, p. 141.

74. C.T. Walker and U. Riege (1979), Compatibility of Actinides with HLW Borosilicate Glass: Solubility and Phase Formation, Ceramics in Nuclear Waste Management, Eds. T.D. Chikalla and J.E. Mendel, CONF-790420, Department of Energy publication, p. 198.

75. R.L. Fleischer, Isotope Disequilibrium of Uranium: Alpha-recoil Damage and Preferential Solution Effects, Science 207, 979 (1980).

76. E.R. Vance, L. Efstathiou, and F.H. Hsu, X-ray and Positron Annihilation Studies of Radiation Damage in Natural Zircons, Radiation Effects 52, 61 (1980).

77. H.D. Schreiber, G.B. Balags, B.J. Williams and S.M. Andrews (1981), Structural and Redox Properties of Uranium in Ca-Mg-Al Silicate Glasses, Scientific Basis for Nuclear Waste Management, Vol. 3, Ed. J.G. Moore, Plenum Press, NY (in press).

78. E.H. Hirsch, A New Irradiation Effect and its Implications for the Disposal of High-level Radioactive Waste, Science 209, 1520 (1980).

79. G.J. McCarthy, High-level Waste Ceramics, Trans. Amer. Nucl. Soc. 23, 168 (1976).

80. G.J. McCarthy, High-level Waste Ceramics: Materials Considerations, Process Simulation and Product Characterization, Nucl. Tech. 32, 92 (1977).

81. J.M. Rusin, R.O. Lokken and J.W. Wald (1979), Characterization and Evaluation of Multibarrier Nuclear Waste Forms, Ceramics in Nuclear Waste Management, Ed. T.O. Chikalla and J.E. Mendel, CONF-790420, Department of Energy publication, p. 66.

82. K.D. Reeve and J.L. Woolfrey, Accelerated Irradiation Testing of Synroc Using Fast Neutrons. I: First Results on Barium Hollandite, Perovskite and Undoped Synroc B, J. Aust. Ceram. Soc. 16, 10 (1980).

83. A.E. Ringwood, S.E. Kesson, N.G. Ware, W. Hibberson, and A. Major, Immobilization of High Level Nuclear Reactor Wastes in SYNROC, Nature 278, 219 (1979).

84. A.E. Ringwood, S.E. Kesson, N.G. Ware, W. Hibberson, and A. Major, The Synroc Process: A Geochemical Approach to Nuclear Waste Immobilization, Geochem. J. 13, 144 (1979).

85. G.J. McCarthy, W.B. White, and D.E. Pfoertsch, Synthesis of Nuclear Waste Monazite, Ideal Actinide Hosts for Geologic Disposal, Mat. Res. Bull. 13, 1239 (1978).

86. D.D. Davis, E.R. Vance and G.J. McCarthy (1981), Crystal Chemistry and Phase Relations in the Synthetic Minerals of Ceramic Waste Forms, Scientific Basis for Nuclear Waste Management, Vol. 3, Ed. J.G. Moore, Plenum Press, NY (in press).

87. D.E. Grandstaff, A Kinetic Study of the Dissolution of Uraninite, Econ. Geol. 71, 1493 (1976).

88. H.D. Holland and L.D. Brush (1978), Uranium Oxides in Ore and Spent Fuels, Proc. Conf. High-Level Radioactive Solid Waste Forms, Ed. L.A. Casey, U.S. Nuclear Regulatory Commission, Denver CO, p. 597.

89. R. Wang and Y.B. Katayama (1981), Probable Leaching Mechanisms for $UO_2$ and Spent Fuel, Scientific Basis for Nuclear Waste Management, Vol. 3, Ed. J.G. Moore, Plenum Press, NY (in press).

90. D.P. Karim, P.P. Pronko, T.L.M. Marcuso, D.J. Lam and A.P. Paulikus (1981), XPS and Ion Beam Scattering Studies of Leaching in Simulated Waste Glass Containing Uranium, Scientific Basis for Nuclear Waste Management, Vol. 3, Ed. J.G. Moore, Plenum Press, NY (in press).

91. A. Ogard, G. Bentley, E. Bryant, C. Duffy, J. Grisham, E. Norris, C. Orth and K. Thomas (1981), Are Solubility Limits of Importance to Leaching?, Scientific Basis for Nuclear Waste Management, Vol. 3, Ed. J.G. Moore, Plenum Press, NY (in press).

92. D. Langmuir, Uranium Solution-Mineral Equilibria at Low Temperatures with Applications to Sedimentary Ore Deposits, Geochim. Cosmochim. Acta 42, 545 (1978).

93. A.E. Ringwood, K.D. Reeve and J. Tewhey (1981), Immobilization of High Level Nuclear Reactor Wastes in Synroc: Current Status, Scientific Basis for Nuclear Waste Management, Vol. 3, Ed. J.G. Moore, Plenum Press, NY (in press).

94. N.E. Bibler and J.A. Kelley (1976), Effect of Internal Alpha Radiation on Borosilicate Glass Containing Simulated Radioactive Waste, USAEC Report DP-MS-75-94, Savannah River Laboratory, E.I. duPont deNemours Co., Aiken, SC.

95. J.A. Stone (1979), Studies of Concrete as a Host for Savannah River Plant Radioactive Waste, Scientific Basis for Nuclear Waste Management, Vol. 1, Ed. G.J. McCarthy, Plenum Press, NY, p. 443.

96. N.E. Bibler (1980), Radiolytic Gas Generation in Concrete Made with Incinerator Ash Containing Transuranic Nuclides, Scientific Basis for Nuclear Waste Management, Vol. 2, Ed. C.J.M. Northrup, Plenum Press, NY, p. 585.

97. J.G. Moore, G.C. Rogers, J.H. Paehler and H.E. Devaney (1979), FUETAP (Formed Under Elevated Temperatures and Pressures) Concretes as Hosts for Radioactive Wastes, Ceramics in Nuclear Waste Management, Ed. T.D. Chikalla and J.E. Mendel, CONF-790420, Department of Energy publication, p. 132.

98. D. Schiferl and R.B. Roof (1978), X-ray Diffraction of Radioactive Materials, Advances in X-ray Analysis, Vol. 22, Ed. G.J. McCarthy, C.S. Barrett, D.E. Leyden, J.B. Newkirk and C.O. Ruud, Plenum Press, NY, p. 31.

99. R.G. Haire and J.R. Peterson (1978), X-ray Powder Diffraction of Einsteinium Compounds, Advances in X-ray Analysis, Vol. 22, Ed. G.J. McCarthy, C.S. Barrett, D.E. Leyden, J.B. Newkirk and C.O. Ruud, Plenum Press, NY, p. 101.

100. E.R. Vance, R. Roy, J.G. Pepin, and D.K. Agrawal, Chemical Mitigation of the Transmutation Problem in Crystalline Nuclear Waste Radiophases, J. Mater. Sci., in press (1981).

101. P.B. Macedo, D.C. Tran, J.H. Simmons, M. Saleh, A. Barkatt, C.J. Simmons, N. Lagakos and E. DeWitt (1979), Porous Glass Matrix Method for Encapsulating High-Level Nuclear Wastes, Ceramics in Nuclear Waste Management, Ed. T.D. Chikalla and J.E. Mendel, CONF-790420, Department of Energy publication, p. 321.

102. J.H. Simmons, A. Barkatt, P.B. Macedo, P.E. Pehrsson, C.J. Simmons, A. Barkatt, D.C. Tran, H. Sutter and M. Saleh (1979), Chemical Durability of Nuclear Waste Glasses, Ceramics in Nuclear Waste Management, Ed. T.D. Chikalla and J.E. Mendel, CONF-790420, Department of Energy publication, p. 263.

103. R. Roy and E.R. Vance, Irradiated and Metamict Materials: Relevance to Radioactive Waste Science, J. Mater. Sci., in press (1981).

104. N.J. Magnani and J.W. Braithwaite (1980), Corrosion-Resistant Metallic Canisters for Nuclear Waste Isolation, Scientific Basis for Nuclear Waste Management, Vol. 2, Ed. C.J.M. Northrup, Plenum Press, NY, p. 377.

105. A.B. Christensen, J.A. DelDebbio, D.A. Knecht and J.E. Tanner (1981), Loading and Leakage of Krypton Immobilized in Zeolites and Glass, Scientific Basis for Nuclear Waste Management, Vol. 3, Ed. J.G. Moore, Plenum Press, NY (in press).

106. R.W. Benedict, A.B. Christensen, J.A. DelDebbio, J.H. Keller and D.A. Knecht (1980), Technical Feasibility of Krypton-85 Storage in Sodalite, Scientific Basis for Nuclear Waste Management, Vol. 2, Ed. C.J.M. Northrup, Plenum Press, NY, p. 369.

107. G.L. Tingey, E.D. McClanahan, M.A. Bayne, W.J. Gray and C.A. Hinman (1980), Krypton-85 Storage in Solid Matrices, Scientific Basis for Nuclear Waste Management, Vol. 2, Ed. C.J.M. Northrup, Plenum Press, NY, p. 361.

108. M.T. Morgan, J.G. Moore, H.E. Devaney, G.C. Rogers, C. Williams, and E. Newman (1979), The Disposal of Iodine-129, Scientific Basis for Nuclear Waste Management, Vol. 1, Ed. G.J. McCarthy, Plenum Press, NY, p. 453.

109. C.F. Smith and J.J. Cohen (1980), Perspectives and Reconciliation of Viewpoints on Risk Assessment Issues, Waste Management '80, Ed. R.G. Post, University of Arizona, p. 151.

110. L.L. Burger (1980), Determining Criteria for the Disposal of Iodine-129, PNL-3496, Battelle, Pacific Northwest Laboratory, Richland, WA.

# Chapter 9.   CANISTERS, OVERPACK AND/OR BACKFILL

### *Synopsis*

*The entire processing of the waste form has to be conducted by remote operation in a 'canyon' facility, and is therefore complex and extremely expensive. However, the next two 'layers' in the multibarrier waste package—the canister and the overpack (backfill)—are both 'cold.' The future potential for increasing the total radionuclide immobilization effectiveness of the waste package is probably much greater in these areas than in the waste form. These topics will also be dealt with in other books in this series and only a brief overview is provided here.*

*Proposed canister materials include metals, ceramics and concrete. Mild steel, stainless steel, titanium alloys and copper are the chief metal candidates. Hot-pressed alumina, alumina-zirconia, and concrete are the non-metallic candidates. Recent work has shown that it is necessary to be sure to adapt the canister material to the particular host rock. Thus titanium is superior to stainless steel in a salt repository. Swedish corrosion studies on both their $Al_2O_3$ and the massive copper canisters have shown that these are more than adequate with respect to possible corrosion, for the 1000-10,000 year requirement.*

*Considerable confusion exists around the terms overpack and backfill. They refer generically to a maximum of two additional layers of materials to provide mechanical and chemical protection. The overpack often refers to a second shell around the canister, made of concrete, a ceramic or some metal. The word backfill suggests crushed rock or fine-grained mineral material to be tamped around the canister. However, the backfill materials with the desired chemistry can be shaped into ceramic "overpack." Besides providing mechanical stabilization of the canister in the hole in the rock, the overpack-backfill has been seen as providing both a material to keep water out (KBS study) a back-up to absorb radionuclides which might leak out from a failed canister, and an additional chemical barrier against "leaching." Proposals for massive cast-iron overpacks do not differ much from making massive copper canisters. Recent work on carefully selecting clays and zeolites to maximize the property to both adsorb and retain the principal radionuclides shows the potential of improving these materials. A novel compositional modification developed by the author is the "doping" of the overpack materials in shaped ceramic overpack or in mineral backfill with other solid mineral sources of the non-radioactive analogues of the primary radionuclides Cs, Sr, An, etc. By making these non-radioactive phases slightly less insoluble than the waste form in the repository ambient, an additional thermodynamic barrier is achieved with relative ease and at low cost.*

# CANISTERS, OVERPACK AND/OR BACKFILL

## Function and Terminology

Many of the terms above are used by different writers to mean different things. Moreover the conceptual framework for the function of these elements of the waste package has not been properly clarified. Finally, the science of materials selection for these functions was virtually totally ignored for thirty years.

In Fig. V our basic waste package figure shows for simplicity only two major layers between the waste form and the lined rock hole. First we have a "canister" which performs the following functions:

    (1)  It is an aid to handling and <u>transportation</u> during processing.
    (2)  It is a <u>mechanical</u> <u>barrier</u> against dispersal and dissolution during <u>transportation</u>.
    (3)  It is a <u>chemical</u> <u>barrier</u> to dissolution after emplacement.

Next we show a layer labelled "overpack." The possible functions of the overpack are:

    (1)  To mechanically stabilize the canister in the hole.
    (2)  To adsorb potentially outward migrating species.
    (3)  To prevent or minimize the egress of water to the canister.
    (4)  To control the chemistry of the waste package most cost effectively, becoming an additional chemical barrier.

As one will see below the function of the overpack may be achieved by an enormous variety of combinations of radically different materials, e.g., cast-iron monoliths weighing 50 tons or micron-size bentonite clays.

In Fig. XII below we show four different combinations of canisters, overpacks and backfills to illustrate the range of options still open. What may immediately strike the reader in the foregoing analysis of functions and in the short sections to follow is the neglect of the engineering of this part of the waste package system. This is all the more extraordinary when one considers that while the actual engineering of the chemistry, crystal structure and microstructure <u>of</u> <u>the</u> <u>waste</u> <u>form</u> must be performed remote in a most expensive "canyon"-type facility, the manufacture and emplacement of these two layers in the system is routine materials engineering. This lack of research effort to date results from the absence of *systems* thinking about the waste package as a whole. In fact, analysis and design of these layers in the waste package provides an opportunity for increasing the immobilization of the radionuclides most cost effectively.

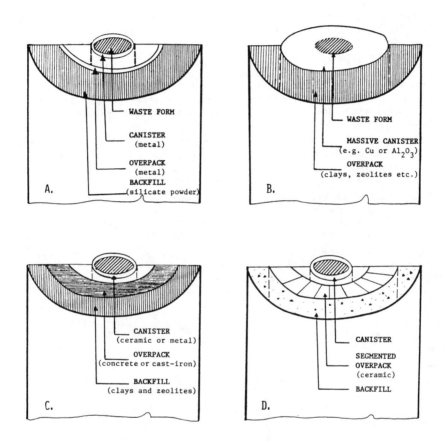

WASTE FORM

CANISTER
(metal)

OVERPACK
(metal)

BACKFILL
(silicate powder)

A.

WASTE FORM

MASSIVE CANISTER
(e.g. Cu or Al$_2$O$_3$)

OVERPACK
(clays, zeolites etc.)

B.

CANISTER
(ceramic or metal)

OVERPACK
(concrete or cast-iron)

BACKFILL
(clays and zeolites)

C.

CANISTER

SEGMENTED
OVERPACK
(ceramic)

BACKFILL

D.

*Fig. XII. Sketches showing many different options for the relative thickness, materials and configurations of canister, overpack and backfill, and the interchangeability of these forms.*

Fig. XII shows four rather different combinations of materials which may serve the functions listed above. *None of these has actually been evaluated beyond the conceptual stage.* Fig. XII(a) shows the kind of "reference" U.S. system which often appears in conceptual designs. It consists of a relatively thin walled metallic canister, followed by a thin metallic overpack, and a backfill of rock fragments and/or sand and clay. Here there has been very little attention to the chemical function of these layers, which serve principally to prop the canister firmly in the lined hole in the rock.

Fig. XII(b) shows a very different option with a massive canister (overpack?) (e.g., 20 cm wall copper) followed by a backfill which has been purposively tailored to exclude water (by the use of quartz and bentonite) maintain a low Eh (by the presence of ferrous phosphate) and possibly adsorb outmigrating radionuclides (by the montmorillonite phases in the bentonite).

Fig. XII(c) shows a third option where the overpack is now a massive layer of metal or concrete, and where the backfill could be a carefully tailored mixture of clays and zeolites which will retain their absorptive and retentive power to the repository temperatures and $H_2O$ pressures.

Fig. XII(d) represents another variant where the overpack is a tailored ceramic material containing Cs, Sr, La, and Au nonradioactive nuclides in ceramic segments or bricks, and where the backfill could be the most logical mechanical-chemical stabilizer of all--concrete.

Clearly therefore there is an enormous opportunity here for materials engineering to optimize this part of the waste package for simultaneous radionuclide immobilization and reasonable cost.

# CANISTERS

In this section we will report principally on the materials which have actually been used to perform only the first of the three functions enumerated above, which the canister will eventually have to perform. We also report on conceptual selection of other canister materials and their projected properties.

The properties demanded of the canister will require tradeoffs between mechanical properties and chemical reactivity properties, availability, cost, and toxicity of the material.

## Materials Proposed

Both metal and ceramic canisters have been used for containing simulated and real wastes in laboratories or pilot plants and are under serious consideration for repository use. Concrete also may be considered seriously.

*Metals:* The "reference" canisters in the U.S. are mild or stainless steel with a relatively high aspect ratio (0.2m x 3m). Such canisters have been used in all the trial "cold" and "hot" melts at the Battelle Pacific Northwest Laboratories—and very recently at the Savannah River Plant. The British and German programs envisage smaller canisters (say, 0.3m diameter x 1.2m) of stainless steel. One version of the Swedish KBS study proposed a massive (20cm wall) copper canister (with, as noted above, essentially an encapsulant equivalent to an inner canister liner of lead). A very large range of other metals has been discussed in conceptual studies, but no actual canisters have been made other than of steel or copper. Of those under discussion, titanium and its alloys has been given high priority by the NAS CRWM Committee (1) and mentioned by the Swedish group (2). At various times it has been proposed to gold-plate or $TiB_2$-coat the metal canisters.

*Ceramics-Concrete:* Concrete canisters of a very wide variety have been used in the nuclear industry for many purposes for a long time. They are standard for low and some intermediate level waste. It is likely that they could be seriously considered for so-called HLW if 40-50 year storage is selected, and a ceramic or cement matrix waste form chosen for ultimate disposal. In the *only* ultimate disposal (in the West) concrete forms both encapsulant and canister in contact with the rock in the Oak Ridge grouting process.

The French program (3) has argued that ceramic containers are the most stable and glass-ceramics, as well as the well-known fusion cast "zirca-frax" ($ZrO_2$-$Al_2O_3$-$SiO_2$) materials are being considered. These are extremely corrosion resistant (even at 1500°C with molten glass). In fact they will function as "overpack" (see Fig. XII(c) and (d)) over the thin steel canisters already being used to accept the glass. The first such containers were to have been made in 1980.

The most striking canister made to date is the 'artificial sapphire' one. The Swedish program has demonstrated experimentally the production of theoretically dense hot-pressed $Al_2O_3$-ceramic (50cm x 3m) canisters. The most impressive feat of this ASEA work--from a materials technology viewpoint-was the hermetic sealing of the two parts of the $Al_2O_3$ canister. In effect, they created an imprevious sealed sapphire geode with a 10cm wall and μm grain size. The corrosion resistance of such materials is of course outstanding in silicate rock environments.

## Properties of Canisters

The conceptualization of what canister properties are meaningful is still inadequate. Certain mechanical (impact) properties are important in the transportation phase--if indeed there is any such phase. Beyond that, the chemical barrier partly governs the total radionuclide release rate. Once in place, the canister becomes part of the waste form-waste package.

Hence "reaction" or "leach" rates become important. Systematic corrosion research under the p, t, and *chemical potential* (i.e., presence of waste and rock) environment of candidate repositories has just begun. What will be valuable would be the analogue of the hydrothermal work on hundreds of waste form samples as a function of p, t, and time conducted by the Penn State group. Thus, one will need to be able to compare stainless steel canister corrosion from 300°C down, with the waste form in a brine saturated repository, with that of waste and copper in contact with basalt. So far, Braithwaite et al. (4) have examined a suite of metals in various rock environments at temperatures up to 250°C; Sundquist (5) has studied glass-ceramics but only up to 100°C. Mattson (6) reports literature values for Ti, and $Al_2O_3$.

In later work Braithwaite et al. (7) have shown the superiority of Ticode-12 (a high Ti alloy) to Stainless Steel 304 for NaCl + $H_2O$-saturated environments; while Rankin (7a) in a 'dry' test found no significant attack between dry Carlsbad salt and 304 Stainless. The closest approach to repository conditions was the work by Westerman (8a) and by Casteels et al. (8b) who simulated the moist-clay environment of the possible repository at Mol and studied some two dozen metals. Westerman found cast iron and copper attacked rapidly in brine but most of the super alloys good in both brine and Hanford for 1000 years. Pitman et al. (8c) also showed the *mechanical* viability of these same super alloys in the Hanford basalt environment, even in a radiation field. The Ti alloys proved to be among the best in Casteels set and the role of $S^{2-}$ and $Cl^-$ in corrosion of the steels was delineated. It is significant that with the exception of the Swedish estimates on the $Al_2O_3$ ceramic canister, no other research has included a ceramic or concrete canister.

Realistically realizable canister lifetimes of hundreds to hundreds of thousand of years have been deduced from these studies. In conducting a rather thorough assessment of the Swedish KBS proposal, the NAS-CRWM (9) committee examined in detail the possibilities of chemical corrosion and mechanical failure, including stress corrosion cracking. It agreed that lifetimes of several hundred thousand years for massive copper in an appropriate overpack in a granite repository are likely. In the light of this consensus and the fact that much of the canister preparation and evaluation is done outside the radiation environment, the evaluation of the waste-forms + canister as a unit will probably become more important (see next chapter).

The mineralogical evidence would suggest that the French and Swedish approaches in using ceramics have much to recommend them from *a very long term view*, certainly in silicate host rock repositories as well as salt. However, processing to make such canisters may be prohibitively expensive. In this regard again making concrete canisters reinforced with non-traditional metals uses a very well-established technology. Such canisters will increasingly become competitive if, for example, defense wastes with their low thermal power are left on site. As we have pointed out in the chapter on waste forms, these hydroxylated materials will show the minimum loss by reaction.

## Resource, Toxicity Considerations

Concrete and ceramics offer one very important advantage--the availability and cost of the raw materials. The toxicity of lead, and no doubt of Cr (in stainless steel) leached in oxidizing environments, would become a significant addition to the health hazard in any failure scenario. Perhaps more significantly, extensive use of say, excessive thicknesses of copper or lead in a major repository would not only be expensive, but as the metal resource availability diminished, it may also prove to be a significant attraction to stimulate human intrusion at a later date.

# OVERPACK-BACKFILL

We have discussed above the confusion with respect to terminology in the context of *a functional assessment* of the different engineered barriers. In fact the inconsistency is far worse than indicated, and reference to many other confusing terms has purposely been omitted. Confusion is caused principally by the inability to distinguish the functions of the barriers. The canister clearly has a unique mechanical-handling function for the waste form. All other barriers between the waste and the rock are placed there as additional *chemical barriers* (except for rough mechanical positioning in a hole). It is this failure to recognize the fact that the waste package is a chemical system (certainly in a failure mode when its components, the radiophases, encapsulant phases, canister, overpack/backfill and intruding solutions will be interactive with each other at the p and t of the repository) that leads to several errors of omission and commission. While we stressed this system-behavior in our earliest conceptualization of the multi-barrier concept in 1973 (Roy, 11a), it was only during the last few years that the experimental data showed that it was this last *barrier which could most easily be used to control the chemistry* of the entire solid phase immobilization system (see Fig. V).

Yet this has not been *utilized* by any groups working in this field. Only H.W. Nesbitt et al. (11b) have very recently shown a comprehension of this important generalization. They state, "In this paper we show that the thermodynamic stability and kinetic reactivity of waste forms will be influenced--*and even controlled*--by the nature of the backfill and repository materials." For years this author has attempted to convey this concept of the waste package as a system (see preface), and Nesbitt et al. agree, continuing as follows: "We emphasize that it is necessary to consider the nuclear waste host, backfill and repository rocks as one *chemical system*" (emphases added).

Thus, a "tailored" overpack can be seen mainly as an interactive chemical barrier within the waste package. The special roles of the overpack-backfill barrier in any repository can be envisaged as follows: (a) to selectively adsorb and fix (retain) radioactive ions, in the case of leakage; (b) to interact with waste to form new crystalline phases under repository conditions; (c) to maintain a reducing environment; (d) to act as a pH buffer; (e) to keep the water from reaching the canister; and (f) to form a thermodynamic gradient of the nonradioactive analogues of the principal radionuclides *toward* the canister.

## Compositions Studies and Their Properties

Although reference has been made to the use of the term 'overpack' to designate a secondary metallic canister, there is not a single published paper where the chemistry of *such* an overpack is discussed. Very recently in the report literature, the team at Westinghouse (10) has proposed the use of massive 25 ton cast iron 'overpacks' around the stainless steel canisters. An evaluation of even the concept awaits more details. Hence we report here on the status of the research on silicate and oxide materials to carry out the first five functions noted above.

The first specifically designed overpack-backfill was that proposed in the Swedish KBS study (2). This consisted of quartz + "bentonite" designed to exclude water and adsorb radionuclide cations possibly leaking from the canister. In addition it contained ferrous phosphate to keep the canister in a relatively reducing environment. Several generic questions can be raised about this particular concept. First, do the base exchange capacities normally measured on clays with cations such as $Na^+$, $K^+$, $Ca^{2+}$, etc. tell us much about $Cs^+$, $Sr^{2+}$, $La^{3+}$, $U^{4+}$ (?), and especially in a multiple cation environment? Second, do the typical partition coefficient "$K_d$" values (Conc. in solid/Conc. in solution) determined at room temperature where only the exchange site ions are involved in the reaction, have any value at all for predicting what will happen at say 200°C–250°C where (a) the clay structure may 'collapse' and lose adsorptivity and (b) the total clay-composition may react in the system (see above).

Some work has been going on in one or two laboratories throughout the world to attempt to provide the scientific background on which greatly improved backfill-overpacks can be designed. First with respect to the unusual sets of ions.

Work by Allard et al. (12) initiated the studies of radionuclide adsorption (including actinides and $I^-$) on clays, more or less under ambient conditions, i.e., under "far-field" conditions and not in the waste package. E.J. Nowak (13) has studied actinide ion $K_d$ on mixtures of hectorite and bentonite in a simulated brine environment at room temperature, and calculated the "breakthrough" times for a one-foot barrier as being in the $10^3$–$10^4$ years. He notes that *prior* heating to 300° did not change the $K_d$ sufficiently. Bird and Lopator (14) have studied the anions $I^-$, $TcO_4^-$ and $SeO_4^{2-}$ again at room temperature, and Brookins (15) has adduced evidence from the clay mineralogy of uranium deposits regarding the immobilization of various ions including uranium.

Komarneni and Roy (16) have explicity determined the adsorptions using *mixtures of adsorbents* (clays, zeolites and oxide gels) and *mixtures of the most important radionuclides*. Furthermore, they have studied the _fixation_ of such ions by the various adsorbers since under certain conditions this retention becomes very significant. They have demonstrated synergistic effects when two absorbers—such as vermiculite and gibbsite—are used with $Cs^+$. They select mordenite and clinoptilolite as the most efficient adsorbing combination.

These studies show that either specific naturally occurring or synthetic clays, minerals, or combinations of clays and zeolites can serve as very effective barriers provided the temperatures are maintained say below 60°C and no lattice interactions occur. They apply with great force to far-field repository conditions where clay beds are involved in the isolation design (as at Mol, or in the USSR). They are also very significant for clean-up situations such as at Three Mile Island in Harrisburg, although no research whatever was used in selecting the adsorbing zeolites in the TMI case. Ringwood (16a) has recently suggested the use of MgO as an overpack instead of the bentonite as used in the KBS study, to react with and keep out water. Others hold that such hydration may set up major stresses around the canister. Ordinary concrete, of course, can achieve the same end since as is not well known ordinary set concrete is only partly hydrated.

However, the influence of temperature and water (i.e., the hydrothermal conditions) on the clays and zeolites both prior to, and during reaction with the canister and the waste form has only very recently been studied in this laboratory. In a series of papers, Komarneni et al. (17) have reported earlier on the influence of high temperatures and hydrothermal conditions on Cs fixation by individual clay minerals and by complex studies. More recently Komarneni and Roy (18) in several papers have reported on reactions between overpack-backfill materials and individual radionuclides on waste forms under various repository-simulating conditions. What became obvious from this work is the fact that the clay minerals are extremely reactive components of this waste-package-chemical-system, and several new phases are formed by these reactions. *Hence the simple model of adsorption-desorption in overpack materials breaks down* if one assumes modest temperatures and water present simultaneously with waste and overpack. These results have been combined by Komarneni and Roy (19) into a summary paper regarding design of overpack-backfill materials.

The most recent innovation in the materials science of the waste package, not previously conceptualized by other workers, was the introduction by Roy (20) of the reverse radionuclide gradient. Fig. XIII below explains the concept:

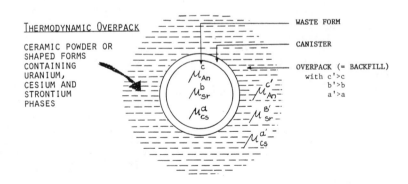

*Fig. XIII. New concept of thermodynamic ore.*

Here the overpack is a "ceramic" material (anhydrous or hydroxylated phases) which generates in the repository environment immediately around the canister a higher chemical potential (= slightly greater solubility) of each of the main species of concern, e.g., Cs, Sr, Ln, An. Thus one creates in effect a thermodynamic barrier to the out-migration (by reaction) of these radionuclides from the waste form. As long as diffusion (as distinct from fluid flow) remains the dominant mode of material transfer the efficacy of this barrier is applicable. Roy has designed several types of physical models and selected various minerals for an overpack which achieves this end. For example, the simplest model simply adds 10% of common minerals such as celestite (Sr), pollucite (Cs), uraninite (An), and various rare earth minerals (Ln) to the silicate overpack clays and zeolites.

In other versions, ceramic overpack segmented bricks (see Fig. XII(d)) containing slightly more soluble phases (such as $CsAlSi_5O_{12}$ and $SrAl_2Si_2O_8$) are envisaged.

## Summary

What is emerging from these recent results is the realization that the canister-overpack-backfill offer a relatively simple means to control the chemistry of the waste-package sytem. Indeed, it has become obvious that a great deal of wasted effort has gone into waste form design without recognizing the impact of the overpack-backfill on the performance of the waste form (see next chapter). Conversely, given the new realization, research on *molecular* engineering of this canister-overpack-backfill barrier is now likely to pay the largest dividends.

# Chapter 10.   EVALUATION OF THE WASTE PACKAGE SYSTEM

### *Synopsis*

*The evaluation of the waste package system has hardly been thought about, let alone carried out. The confusion of the waste package as essentially equivalent to the waste form has been a grossly misleading error. The second egregious error indulged in, ad nauseam, is the "leach test" as a means of supposedly ranking the performance of waste forms, and by implication waste packages. Insofar as the waste package after emplacement in a repository becomes a system in the thermodynamic, chemical sense, it can only be evaluated as a system. Component evaluation is without meaning for the whole system even if the test itself were significant. The errors and dangers in the typical leach tests are delineated.*

*Alternative tests called repository-simulating-tests (RST) involving the determination of the quantitative changes in the total solid and liquid "reactants" as they transform to "products" within the waste-overpack-rock-metal-solution system at temperatures and pressures bracketing possible values for commercial and defense waste repositories are described.*

*A comparative summary of the physical and chemical properties of various waste forms is presented to be used where appropriate within the context of the RST.*

*The final section is devoted to a few high probability scenarios of the actual technology which may be deployed in the U.S. for defense wastes and separately for commercial wastes at a much later date, and the impact which these scenarios will have on the waste package.*

# EVALUATION OF THE WASTE-PACKAGE SYSTEM

## Difficulties and Limitations

This chapter is a short but necessarily general summary on the topic of evaluation of performance in order to complete the presentation on the waste package. This topic is treated more fully in another book in this series.

In the preface to this book we have repeated our long-standing emphasis on the fact that the "waste package" or more accurately the solid-phase immobilization subsystem is a major "subsystem" of the total waste disposal system. The other major equivalent unit is the "geologic isolation" subsystem. While there is of course a systems-interaction between these major components they are *not* part of the same *chemical thermodynamic system*, and hence they can be isolated for study and separate evaluation. But what is needed urgently is an experimental means to evaluate the performance of the waste package subsystem considered as a single-point source of radionuclides that will be released as a function of time especially in the $10^3$-$10^5$ years range.

Up to now 99+% of the evaluation work has been performed *on the waste form alone*. This has naturally led to results which cannot be used in any meaningful way.

In thermodynamics a system is "that portion of the universe isolated for study." The key work, "isolation," refers to the fact that no external variables: composition, temperature, pressure, etc. can be changed from outside the system. It is at this point that experimental work on the attempted evaluation of the waste package as a whole must justify itself. First it is obvious that under any conditions projected for a repository, the waste package (with a volume of some tens of $m^3$) will not interact as a whole, or become part of the same chemical system as the entire repository (volume $10^8$-$10^9$ $m^3$). There is no way for the compositions of waste package and total geological host to mix and interact. There may be very partial reactions between any material emerging from the waste package, and the inside surfaces of faults, fissures and cracks along which *fluid* flow may occur. Furthermore, the temperatures involved (25-50°C) throughout the mass of the host rock make impossible any substantial compositional exchange via either dissolution or solid state reaction. The boundary conditions assumed in this argument may of course be violated in very low probability scenarios. For the sake of completeness, I enumerate below in (qualitatively estimated) decreasing order of probability the circumstances which are *excluded* from the argumentation which follows in this chapter.

1.  Effects of nuclear weapon explosion on waste package, or repository.

    While those who have seriously analyzed the dangers of nuclear war
have alluded to the threats involved, no quantitative analyses are known to
the author which evaluate the probabilities and the effects of purposive
attack on a radioactive waste-facility or repository.  Since many studies
such as those at the Max Planck Institute by Weizsacker and co-workers (20)
indicate such a high probability *of nuclear war* (∿0.50 by the year 2000),
one needs only to determine where waste facilities are ranked in the order
of targeting options, to complete risk analysis scenarios of radionuclide
dispersal consequent upon nuclear war.  The radionuclide content in a
reactor storage pool or reprocessing plant with a large away-from-reactor
storage pool, would be a very substantial addition to the total fallout
from even a 1 megaton weapon.  Waste package performance will not likely
have much effect on the total effects from a direct hit, although surface,
land-mine and subseabed *repositories* would perform very differently.  In
any event, it is interesting to note that such a high probability risk of
(additional) radiation exposure from wastes has been ignored so far.

2.  Nuclear excursions or accidents in the waste processing or repository step.

    Here again the probabilities of such incidents, especially if one
considers the disposal of spent fuel, but in any case during reprocessing
and storage are clearly substantial compared to other accident scenarios.
Indeed, the largest nuclear accident in history which occurred at Kyshtrym
in the USSR in 1967 has often been attributed to the back end of the fuel
cycle.

3.  Major transportation accident or hyjacking during shipment of waste.

    Such an accident or threat to disperse by conventional explosion the
contents of a waste shipment could expose the waste form (or spent fuel) to
a very different chemical environment than the repository without any other
components of the waste package.  Hence its performance would be evaluated
very differently.

4.  Major earthquake or unexpected tectonic activity causing drastic change
in hydrology.

    Since the repository will certainly be selected with extreme care to
avoid such possibilities, such events must be taken as very low probability
events (several orders of magnitude lower than say the nuclear war scenario
listed above).  Moreover, the major threat from such an event will be the
access of very large volumes of water through a major fault or fissure
right through the repository.  It is only in this extremely low probability,
series of events that some (*though not all*) of the conditions roughly
simulated in a leach test are approached.


Goal of Waste Package Evaluation

    While it may be important to evaluate the performance of the waste
package under the low probability eventualities listed above, it is
essential to evaluate its performance under *the most likely situations*
under which it will be required to perform.  These conditions are dictated

by two wholly unrelated systems:  (a) the storage subsystem, i.e., how old
the solidified waste will be and (b) the repository design.

While waste form and package design can be changed to accommodate
these variables, any evaluation can therefore only be situation specific to
the age ($\sim$ radionuclide content $\sim$ the temperature) the waste package and
the host rock.

While we may not therefore be ready, at this stage of our knowledge,
to evaluate the total package *system*, we can at least describe the most
significant properties of each component of the system.  We start with the
waste form.


## Properties and Processing of Waste Forms: A Comparative Summary

This section presents a comparison of the product and process
characterization of each of the major candidate waste forms.  It reaches
the highly significant conclusion that especially at low temperatures (say
100°C) the product properties are *relatively* close to each other for well-
made samples of the three major waste forms, a borosilicate glass, a dense
ceramic and a concrete-encapsulated dilute waste.  Hence the choice will
likely be made on the basis of processing differences.

In order to give the general reader who wishes to compare the various
candidate forms with respect to any parameters a compact reference source,
we have brought together in Table 23 the *data* on the various products and
processes in the several waste packages which are possible.  No effort is
made to 'rank' any forms, rather the aim is to provide the information on
which the *present state* of the development of any waste package can be
judged.  We emphasize the following points:

1.   There are three major materials-classes of waste forms:  monolithic
glass, consolidated ceramics, and cement-matrix materials.  All three major
waste forms are sufficiently well developed in the engineering sense *that
with the lead time available,* and no further discovery, they could be
brought up to full-scale waste operation in a suitable waste package.

2.   It is highly unlikely that the same SPI subsystem, or waste
package, or waste form will be optimum for the wide range of different
disposal problems at hand.  There is compelling technological necessity to
develop a *small* number of rather different SPI systems.  For example, it is
patently unlikely that the same waste form will be optimum for commercial
wastes and for the 100x more dilute and chemically radically different
defense wastes.  The nuclear power enterprise, which is caught up in its
own special socio-political situation, cannot risk being caught with a
failure by using a single approach to waste solidification.

3.   No waste form has a dominant position in the field in all respects.
It is certain that no waste package will be the best in all respects--
stability and insolubility of the product, process simplicity, and cost.

4.   In the laboratory, one has already achieved essentially a very low
release waste package for any seabed or geologic repository disposal in the

TABLE 23

Comparative Data Waste Forms

| Generic Waste Form | | Description | | Product Characteristics | | | Process characteristics | | Stage of Development | | | Advantages/Disadvantages |
|---|---|---|---|---|---|---|---|---|---|---|---|---|
| | U.S. Ref. | Radio Phase(s) | Encapsulant Phases | Solubility | Dispersibility | Reactivity C H₂O @,t | Temp. Press. | Complexity | Res. Data in open Literature | Tech. Develop. | Hot Engr. Exper. | |
| GLASS | a | Borosilicate glass (+RuO₂ + Pd) | None | good | excel. | fair | 1300°C | High Mod. | v. high | High | High | Momentum for this system / *Highest Temperature Process* |
| | b | varying chemistry & waste content | None | improved by dilution | excel. | improved by dilution | 1300°C | High Mod. | v. high | High | High | Total systems must be compared if dilution is utilized |
| CERAMICS | a | Set of "insoluble" phases. Monazite, pollucite, spinel, perovskite, zirconolite, etc. | None | excel. | v. good | v. good | 1200°C | High Mod. | v. high | Mod. | v. low | very good product / *Needs Process Development* |
| | b | Hollandite, zirconolite, perovskite | None | excel. | v. good | excel. | 1200°C 5 kbar | High | Low | Low | None | Potentially fine product if realistic waste loading possible / *Process Development* |
| | c | CsCl, SrF₂ (SrTiO₃) | (Hastelloy) + stainless) | poor (excellent if can included) | fair (Excel. if can inc.) | poor (excellent if can included) | 500°C | Mod. | Low | High | v. high | For partitioned wastes System completely engineered and working / *Radiophase not optimum* |
| CEMENT | a | Largely unknown | C₂S, C₃A, CSH-I | good | v. good | fair | Room Temp. | Low | High | Mod. | Low | Extreme simplicity, low cost, raw materials |
| | b | Largely unknown | C₂S, C₃A, CSH-I | good | excel. | fair | ~200°C s.v.p. | Mod. | Low | High | None | Low temperature process, etc. |
| METAL MATRIX (Mol) | | Phosphate (or borosilicate glass) | Lead | excel. | excel. | excel. | 1250°C 400°C | High Mod. | High | High | Mod. | *slightly More Complex Process;* excellent product; *Wasteful of Lead; Toxicity of Lead* |
| CERAMIC MATRIX | | SrTiO₃, Zirconolite, Zeolite, etc. | TiO₂ | v. good | excel. | excel. | 1000°C 1 kbar | High | Low | Mixed | v. little | Low temperature first stage; needs process development |
| CEMENT-MINERAL MATRIX | | Largely unknown | C₂S, CSH-I, etc. | fair | maximum | fair | Room Temp. | Low | v. high | v. high max. | | Minimum cost, max. system safety |
| CERAMIC MATRIX | | Largely unknown | Al₂O₃ spinel | v. good | v. good | v. good | 1000°C <1 kbar | High-mod. | v. low | low | same | Process line shortened as compared to glass; low temperature first stage |
| METAL MATRIX | | Largely unknown | Copper (Fe) | v. good | excel. | v. good | Room Temp. | Mod. | v.v.low | low | none | Lowest temperature alternative to concrete |
| GLASS-MATRIX | (a) | tailored ceramics see above | SiO₂-glass | v. good | v. good | v. good | | | | | | Too little data for serious evaluation |
| | (b) | waste slurry | high-silica glass | v. good | good | v. good | | | | | | |

HIGH-CONCENTRATION FORMS

LOW-CONCENTRATION FORMS

upper 1000 meters of the earth. Coated ceramics in a metal matrix are so good that only the composition and the coating can be improved slightly, as a stable waste form. Yet this is, of course, the Cadillac of waste forms and probably not justified except for fresh commercial wastes.

5. Process simplicity will probably dominate over product properties *since the latter cluster rather closely for most waste forms*. Encapsulation strategies appear to offer many advantages. Furthermore, room or low temperature processing offers many maintenance advantages over the very high temperature processes required for glass and ceramics. Research on the latter has received very little support so far so that comparison of process effectiveness etc. are impossible at this stage.

The last point in the summary above needs to be expanded upon. We have noted in Chapter 9 that, to date, the canister, overpack and backfill part of the waste package system have received very little attention. In this connection of summarizing the status of knowledge about the waste form, it is evident that the *research base on the process to make the waste form* has received vastly less attention than the properties of the product being made. Yet because the waste form is but one part of the chemistry of the system, its properties are in fact neither determinative of the final system's properties nor are they of any intrinsic importance. On the other hand, the processing step of the waste form is by far the most important *and unavoidable* single hazard to real workers. Innovation in simplification and increasing the reliability of the process would therefore pay enormous dividends both in safety and costs. Alternative processing studies are an obvious high-priority research goal.

In the foregoing section we reported on the properties of the various candidate waste forms. We were careful not to imply *any* connection between these properties and the evaluation of the performance of the waste form or waste package. Indeed, in Burkholder's analysis (21) of the total system performance, the properties of the waste form do not even appear. We turn now to that task, first recognizing that evaluation of any total radioactive waste management system requires much more than the technological functions. To set that evaluation in perspective, we show its place in the decision tree in Fig. XIV below.

We note first that the technological system is a minor part of the total matrix and the contribution of scientific or technological advances will be modest, except *insofar as they radically change the cost*. Secondly, we note that the technological effectiveness is a series addition of two terms, of which the total waste package is one. The waste package as a unit has to be evaluated for its sole function-controlling the radionuclide release rate *under the given repository conditions*. This must form the principal goal of all evaluation research.

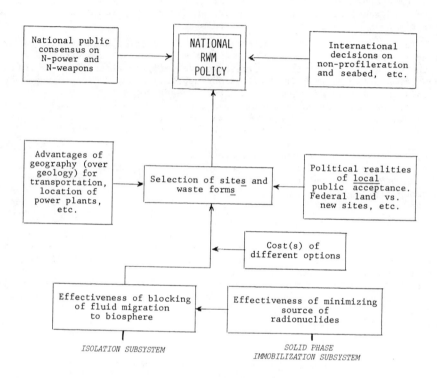

Fig. XIV. The factors involved in the RWM decision tree. This shows the relatively minor role for the technology in deciding where to place the wastes. Geography, not geology, will control site selection, and from it will be determined the rock choice and the waste form choice.

# WASTE PACKAGE EVALUATION
# LEACH TESTS: WRONG TARGET; WRONG CONDITIONS

In the attempt to evaluate the waste package, three very serious scientific errors were committed, starting in the early fifties, and persisting to this day. The first error is to equate the waste form with the waste package. A glance at Figs. V, XII and the figures below will show that this is a ludicrous assumption, namely: that by measuring the radio-nuclide release from the waste form, one will be able to deduce the radio-nuclide release from the waste package. The second major error was a misunderstanding of the relation of thermodynamics and kinetics. Tests were devised under conditions which were easy to accomplish in the laboratory-e.g., at temperatures from 25-90°C rather than those which were expected in the repository (up until the mid-seventies, this design temperature was 300-500°C). It was then expected that by measuring at a few temperatures in the range 25-90°C, one could extrapolate to higher temperatures. The error, of course, is that the reaction(s) which occur(s) at the higher temperatures may well be completely different from those which occur at low temperatures. In this event, absolutely nothing can be learned about the high temperature processes by studying in great detail the lower temperature processes.

As the multibarrier nature of the waste package itself became more widely understood, a third error was introduced. This was the concept of the "linear sum" of radionuclide release. According to this concept, the waste form, canister, overpack, rock etc. do not interact at all as a system (in other words, there is no simultaneous chemical reaction between, say, waste form, solution and overpack). Hence one could study waste form dissolution, then corrosion of canisters, then overpack interaction with solution etc. The models were very naive. For example, the overpack adsorption of ions putatively released from the waste form were calculated, but never the reaction of the ions released from the overpack with the waste form.

Since the middle seventies, work in this laboratory has in fact radically transformed the concept behind testing of the waste package. Before dealing with the practice and results of such alternative tests, we need to look in more detail at the so-called "leach tests" as they are performed.

## Previous and Current Practice of the Leach Tests; and Some Results

Although there are a large number of variations, essentially leach testing consists of soaking either powder or a solid specimen in water--either distilled or deionized-at temperatures between 25°C and the boiling point.  Rymochowicz (22) lists thirteen individual variations of leach test arrangements that have been used in the U.S., Canada, and Europe.  In 1978-79 in an attempt to standardize leach testing practices, the Materials Characterization Center established by the DOE set up, tentatively, a new standard "leach test."  This test requires the following procedure:

> "Monolithic" specimens are suspended by teflon monofilments into a leachate at 90°C in a static environment.  The ratio of geometric surface area to specimen volume is maintained at $1 \times 10^{-2}$ mm$^{-1}$.  One leach test container is to be used for each test.  The resulting leachates are to be analyzed and the solution analyses reported as normalized leach rates at a given test time in g/m$^2$.

A very large body of data exists (which we will not refer to in detail) on the leach rates of borosilicate and phosphate waste glasses,* the better materials yielding cesium or other alkali leach rates of $1 \times 10^{-5}$ to $1 \times 10^{-7}$ grams per sq.cm. per day (23,24).  *Comparably* detailed data-in regard to concentration of radionuclides, exposure of radiophases, etc., have not been reported for *all* alternate waste forms.  However, many of the data reported on the more developed ceramic forms such as the tailored ceramics of McCarthy and Ringwood and some cement forms all fall in the range from $1 \times 10^{-4}$ to $1 \times 10^{-8}$ gm/cm$^2$ (25,26,27).  This is an important and key finding of the "leach rate" data.  The scatter in the data within a laboratory due to procedure and sample to sample variation is not less than ± a factor of 10.  Including different samples made by and measured by different laboratories, the scatter is at best between $10^1$ and $10^2$.  But the leach rate differences between a good cement encapsulant sample, a good glass, and a good polycrystalline ceramic is probably within the $\pm 10^2$ scatter, although the order of increase of the dissolution rate is probably in the order above, under the artificial conditions of the test.

From dissolution rate measurements, a certain amount of knowledge has been accumulated on the *mechanisms of the leaching* reactions.

The leach rate of waste glasses typically decreases with time and increases with temperature.  The fall-off in the leach rate of glasses is assumed to be due to the buildup of surface barrier layers, but few detailed characterizations of waste form surfaces have been made.  Analysis of the outer few micrometers of leached simulated glasses (24) revealed a complex superposition of inward-diffusing hydrogen, outward-diffusing alkali and alkaline earth metals, and a buildup of barrier layers of some rare earths and transition metals.  For the same reaction, the leach rate appears to increase exponentially with temperature and probably exhibits Arrhenius behavior--that is, a linear relationship between the logarithm of the leach rate and the reciprocal of the absolute temperature.

Analogous studies on the kinetics and *mechanisms* of dissolution of the principal ceramic radiophases and on ceramic aggregates have been underway in our laboratories, under the direction of W.B. White and his colleagues (27).  These data show for example the very slow dissolution rate of pollucite and the even slower rate of monazite, and also the major changes

---

*See Appendix 1 for some of the details.

which can occur with changes of pH (in the case of pollucite, the
dissolution rate is closely proportional to $H_2^+$ activity).

In exactly analogous vein to the pH influence, the oxidation potential,
Eh, of leaching solutions influences the rate of dissolution of variable
valence elements. As a general rule, for the heavier elements, higher
oxidation potentials yield more soluble species. Examples include uranium
(and the transuranics) in which $U^{6+}$ compounds are far more soluble than $U^{4+}$
compounds, and the technetium in which $Tc^{7+}$ compounds are both soluble and
volatile, whereas $Tc^{4+}$ compounds are refractory and probably very insoluble.
Few dissolution measurements of waste packages have taken account of
oxidation potential and little is known about the *rates* at which, for
example, the uranium in spent fuel, would oxidize to a more soluble form,
although there is a great deal of data on $UO_2$ stability as a function of pH
and Eh.

## Conceptual Errors in Using the Leach Test to "Evaluate the Waste Form"

The most important of the errors has already been pointed out above.
One *cannot meaningfully* evaluate the waste form outside the specific waste
package system in which it is embedded. If the waste, canister, overpack
and/or host rock interact, *there is no direct connection between the
leaching data* obtained in de-ionized water, and those which would be
obtained in the chemical potential dictated by some real waste *package*.

However, the errors do not end there. Certainly any evaluation of the
waste package should consider both the normally functioning mode and
failure modes. It is clear that we must consider the most probable failure
modes as well as the less likely ones. In Fig. XV below we illustrate the
two key variables involved in a repository failure. The first is the
solution:solid ratio during the time of reaction, and the second is the
ratio of the time for transit of a volume of solution *through* the waste
package compared to the time for equilibration of the solution with the
solid phases of the waste form, canister, overpack and rock.

The much more probable failure mode is one in which relatively small
amounts of liquid breach the waste package, and the transit through the
package is relatively slow (compared to the equilibration time). This
means that an evaluation of the waste package for the high probability
failure defines two experimental conditions:

    a.   a low water:solid ratio.

    b.   an essentially static or closed system test, since equilibrium is
obtained between the solution and all the different solid phases of the
waste package.

Using the same argument, containment failure is most likely when the
temperatures in the repository are highest. This temperature can of course
be adjusted by design-with considerable impact on cost and repository
size--but it seems to vary from 100°-450°C for various *currently* proposed
national systems. Hence it is clear that waste package evaluation should
be carried out at these temperatures (and pressures).

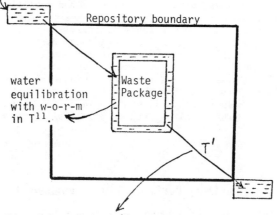

Volume of water transiting through
waste package/unit time (determines
realistic water/solid ratio in test).

Repository boundary

water
equilibration
with w-o-r-m
in $T^{11}$.

Waste
Package

$T'$

Time for transit ($T^1$) through package compared
to time ($T^{11}$) for equilibriation with solids
determines open or closed system nature of test.

*Fig. XV. Shows components of realistic waste repository and the actual
solid phase immobilization system which is an interactive set consisting
of waste-overpack-rock-metal-solution (w-o-r-m-s). The most probable
failure scenarios are those in which $T^1>T^{11}$ for the solution transit
through the waste package compared to the equilibriation time. $T^{11}$ from
lab experiments at temperatures of say 200°C is in the order of months.*

c.  at modest temperatures and moderate water pressures:

Fig. XVI below shows that the "leach test" conditions fail to simulate
all three of the most significant parameters which would describe the
environment of the repository in the most probable failure mode.

The question may therefore be raised:  Are there any conditions where
the leach test data may apply?  One less likely failure could be a great
increase in the volume of fluid phase, and more rapid transit of the
fluid.  Indeed the more rapid transit will lower the temperature somewhat.
But even in such a case, while the direction of change from the repository
simulation will be *towards* the leach test conditions, it will not remotely
approach those conditions.

The more significant question is:  Are the leach test data merely
meaningless or can they be positively misleading?  Here the new results
coming from the new repository-simulating tests begin to speak for
themselves.  The use of leach data to rank waste forms as a means of
qualifying or rating waste forms or waste package is grossly misleading and
hence dangerous nonsense.  The temperature, the pH, the overpack composition

| Repository Conditions | Leach Test Conditions |
|---|---|
| 1 <u>Temperatures</u> will range <u>minimum</u> of 100°C for U.S. defense wastes to up to 300-400°C in <u>some</u> power-waste repositories (e.g., Japan) and <u>pressures</u> to 300 bars | <u>Temperatures</u> range from 25-100°C; 1 atm. pressure |
| 2 Water/solid ratio in most likely failures is << 1 | 3 Water/solid ratio > 1 |
| 3 Ionic activity high; determined by backfill (and rock and canister) | 4 Deionized water used to date |

*Fig. XVI. Major differences between repository thermodynamic conditions (t,x,p) and those in a leach test, making the data from the latter irrelevant to the former.*

can completely invert the order of radionuclide release from two different waste forms. Here the simple-minded use of leach data (with the limitations already noted) can therefore mislead national policy by suggesting that a particular waste form may be more "leach resistant" than another, when in fact the reverse may be true, or there may be no differences whatsoever. Given the nature of the typical waste form, especially the glasses, and the typical overpack and rock, the leach test is almost certain to overstate by some orders of magnitude the source term of ions supplied for migration. *Hence it is the more cost-effective, simpler and safer-to-manufacture waste forms which will be screened out by this misleading procedure.*

## The Alternative: The Repository Simulation Test

The reasoning developed above suggests that to evaluate meaningfully the waste package performance, at least under the most probable failure conditions, any test should simulate the repository with respect to the following parameters.

a. temperature (and pressure)

b. chemical potential of all species caused by the rock, canister, overpack

c. water:solid ratio, and closed–open system

Having been involved since 1948 in experimental petrology research to simulate the p,t conditions of the earth's crust at various depths, it was natural that The Pennsylvania State University group introduced *hydrothermal experimentation* to the radioactive waste management community for evaluating waste forms (28). These hydrothermal studies simulated the (a) condition much better than the tests. Moreover, in the very first paper on hydro-thermal waste form reactions, we explicitly recognized the influence the host rock would have on the reactions. We next introduced the practice of incorporating *waste-rock interaction* under repository p,t conditions as a means of waste package evaluations (29). Over the next few years, experi-ments, numbering in the several thousands, were performed to evaluate the stability of waste forms using methodologies developed in these laboratories and at the Carnegie Institution in Washington over the last three decades (30). Pieces of the waste form and the host rock were *sealed* into noble metal tubes with *small* amounts of water and held for periods of time up to 1 and 12 months, which our experience has shown provides a not unreasonable simulation of reactions found to occur in geological times. (The lower the temperature, the less reliable this approximation.) Thus these experiments, in our judgment, most closely simulated the effects likely to be encountered in a repository failure--low water:solid ratios and an effectively closed system. We then first extended this work (31) to the open system simulation involving the rock and the waste form. These flow-through experiments where water saturated in various rocks at elevated p,t was allowed to react with the waste form, at elevated p,t. The most recent test put into place has been as a result of the work we have done on overpack or backfill reactions. This overpack material *now appears to be the most significant part of the* chemical environment seen by the waste form, and it must be included as part of the closed (or open) system being evaluated. The overpack (backfill) is also by far the most easily "engineered" of the barriers which will influence the chemical environment of the repository. Hence the backfill-solution-waste reactions will dominate the repository simulation environment, except in the case of rock salt (where the backfill will still ve very influential). Next, it was shown that while *the canister may be quite inert* it must also be included in the experiment since it does react differently in different environments.

The tests of the waste package system which at present appear to be the most useful in providing the best experimental simulation of the waste package-in-repository environment are as follows:

1. A physical assemblage not unlike a scaled-down version of waste package: Surrounding a pellet of waste form is a series of concentric layers of canister metal, overpack, and host rock; this assemblage is sealed into a gold or teflon capsule with a small amount of water (see Fig. XVII).

2. The runs are made at modest $H_2O$ pressures (100-300 bars) at temperatures from 100-300°C.

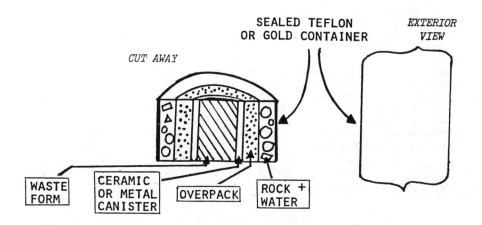

SEALED TEFLON
OR GOLD CONTAINER

*EXTERIOR
VIEW*

*CUT AWAY*

WASTE
FORM

CERAMIC
OR METAL
CANISTER

OVERPACK

ROCK +
WATER

*Fig. XVII. Laboratory experimental set up for sample containers for
hydrothermal repository simulating tests (RST) of waste packages.
It consists of concentric nesting pellets of waste form canisters,
overpack and the host rock.*

    3. After cooling, the elements *in solution* are analyzed, and the
chemical and structural changes in each of the waste-package layers
determined by diffraction and electron (or ion) probe analyses.

    4. An open system test with the rock-saturated groundwater extremely
slowly entering and leaving the chamber with the simulated waste package
will also be significant-although none of these have been carried out so
far.

    This scale of test provides a workable model test for large numbers
of experiments which are necessary in order to test the large number of
variables: waste forms, canisters, overpack, pressure and temperature on
the chemical equilibria. At a later date, larger scale tests for evaluating
mechanical failures can be made on a small number of preferred assemblages.

## Experimental Results of R.S.T.

    The potential of "engineering" the overpack materials *to control* the
chemical environment has recently been recognized by a few groups. Thus

Apps and Cook (32) state, "The control exercised over engineered barriers should resolve the questions of variability and uncertainty, *so that these will be less than those inherent in the geologic media*." Even more explicitly, H.W. Nesbitt et al. (33) state, "In this paper we show that the thermodynamic stability and kinetic reactivity of waste forms will be influenced--*and even controlled*--by the nature of the backfill and repository materials" (emphasis added). Their experiments show the greater stability of sphene over perovskite in a waste form + clay + granite system. Extensive studies by Freeborn and White (34) and Scheetz and White (35) in this laboratory have shown the major differences made by the host rock material on the nuclides taken into solution from a variety of waste forms, in different waste-rock-solution configurations. Similar studies in sealed teflon containers which include the metal canisters show remarkable differences in corrosion rates of the canister candidates depending on the same rock or waste variables, showing that the *linear addition* concept of the 'corrosion' or 'leaching' of each of the components is totally untenable. Table 24 is one summary of such evaluation. Recently, Sasaki, Komarneni, Scheetz and Roy (36) have specifically demonstrated the major impact of the backfill on the reaction and dissolution of source ceramic waste forms. Although unable to provide the basis for a quantitative estimate of release from a specified waste package, these results constitute more than sufficient evidence to demonstrate that "leach tests" must be abandoned now and be replaced by such repository simulating tests, in order that we may start accumulating data for evaluating the waste package subsystem's effectiveness, for the first time.

### TABLE 24

#### EXAMPLES OF REPOSITORY SIMULATION TESTS

*Variables studies in 5-dimensional matrix:*

| | | |
|---|---|---|
| 1) | Waste forms: | 76-68 glass; ceramic; spent fuel; [blank] |
| 2) | Canister material: | 304L stainless; zircaloy; copper; mild steel |
| 3) | Overpack: | shales -- carbonate-rich; clay-rich; reduced |
| 4) | Fluids: | deionized water; saturated NaCl; connate groundwater, bittern brine |
| 5) | Temperature @ 30MPa: | 150° and 250°C |

*Analyses Mode of 24 Elements in Solution*
*Examples of data (illustrative elements only)*

| | Effect of canister | | Effect of fluid | | Effect of host-rock | | |
|---|---|---|---|---|---|---|---|
| | 150°C, 1 mo., 300 bars. Salona shale, connate waste 76-68 glass | | 250°C, 1 mo., 300 bars Antrim shale, 76-68 glass | | 150°C, 1 mo., 300 bars Cu canister, connate fluid, 76-68 glass | | |
| | SS 304L | Copper | Connate $H_2O$ | Deionized $H_2O$ | Antrim reduced shale | Salona (calcite) shale | Braillier |
| B | 1750 | 500 | 1250 | 1510 | 320 | 500 | 770 |
| Cs | 10 | 15 | 4 | 5 | 20 | 15 | 5 |
| Cu | - | <5 | <3 | <3 | <5 | <5 | 20 |
| Mo | 1050 | 400 | 610 | 800 | 430 | 400 | 600 |
| Na | 4300 | 4450 | 6610 | 6500 | 4650 | 4450 | 4750 |
| U ppm | 560 | 410 | 250 | 290 | 260 | 410 | 1090 |
| Sr | <5 | <5 | 3 | 3 | <5 | <5 | <5 |

# SOME PROBABLE SCENARIOS FOR RADIOACTIVE
# WASTE DISPOSAL IN THE UNITED STATES

The history of nuclear power has been replete with unexpected surprises. The peaceful atom once, in all good faith, expected to make electricity too cheap to meter has become a minor actor in the worldwide energy picture. The public antipathy, for a complex set of reasons, to nuclear power could not have been predicted in the fifties. The rise of the environmental movement quite independent of nuclear power issues was likewise unpredictable. Nor could one have predicted that neither of these factors but the high interest rates would seriously compromise the once-rosy future predicted for nuclear power. The overriding factor-the softening of the demand for electricity (below highly inflated projected values)--could have been, and was, predicted by the wisest among us, like the economist E.F. Schumacher. While the nuclear waste issue does not appear in this list of most important reasons for the slow growth of nuclear power, it has played a significant role in the debate on nuclear power. "Predictions" about the future of nuclear wastes are therefore subject to some uncertainties, but in some ways, the bandwidth of uncertainty is less than that for nuclear power. Some states have passed lows that put a hold on nuclear power development till the nuclear waste issue is "solved." This tends to put some pressure on the government to "solve the problem." Although the public's opinions are shaped in deep ignorance of many basic facts, the leadership has slowly become aware of many very significant highlights:

- There is no production at all of power plant reprocessed wastes (and virtually no stock) and it is unlikely that there will be for the next 15-20 years. If the reprocessed wastes of the year 2000 are held for 40-50 years before disposal, we have at least 50 years to develop our best approaches.

- Virtually all liquid waste comes from defense operations and all such wastes are stored on very large federal reservations.

- Nuclear wastes have been and are being *solidified and disposed* of on U.S. territory every day (at Oak Ridge).

- *Vastly more nuclear waste has been solidified in the U.S.* than in France, and the technology is ten years old, simple and in hand. Some 500 megacuries of such solidified waste exists at Hanford and Idaho.

- The enormous cost in terms of dollars, transportation hazards, social disruption due to polarization of the population on the nuclear issue is only very slowly being realized. This can change but only very slowly.

- Hence the greatest danger in the nuclear waste management is a completely unnecessary sense of urgency.

In the light of this, one may well speculate what *sociopolitically dominated decisions* are likely to be made and what impact they will have on the technological developments and choices described in this book. It is certain that the sociopolitical decisions will impact (probably indeed completely dominate) the technological areas and not vice versa as has been true in most other high technology spheres. Important examples of such factors are:

1. Split jurisdiction over nuclear waste materials in the many committees of the U.S. Congress will lead to minimum change solutions.

2. Political desire by the Armed Services Committee to permit minimum "interference" on defense wastes will lead to a series of choices adversely affecting technological optimization of the waste package or the isolation for commercial wastes.

3. Realization of the magnitude of the costs will force the choice on defense wastes towards continued deferral, and eventually shift the choices and research and development towards much lower cost options.

4. Since international competition is not involved, grossly inefficient and overpriced national solutions are not ruled out, except by a vigilant public.

5. Historical accident places the interest of the extreme wing of the environmentalist movement in total synergesis with the traditionalist industrial establishment: both are working arduously for very different reasons for the *maximum cost solution* with little concern for cost effectiveness.

### Probable U.S. Waste Disposal Scenarios

On the basis of this analysis of the sociopolitical forces actually operating, one could visualize the following scenarios for actual waste disposal in the U.S. in descending order of probability.

### A. Hanford wastes solidified and disposed of in present tanks starting in the 1990's.

Wastes at Hanford will be solidified inside the tanks by use of tailored concrete added into the tanks with or without any on-site out of tank treatment. The steel tanks full of solid concrete will be then entombed in concrete or soilcrete a few feet thick. This solution requires no transportation, keeps all operations on federal land, and possibly avoids NRC jurisdiction, and involves the minimum cost and at least the maximum near-term safety.

B. Savannah River Wastes emplaced in caverns excavated in bedrock below plant by grouting procedure.

This option uses the same reasoning as at Hanford with regard to the geographic imperative. Shipping such large volumes of waste out of a federal repository across dozens of states to some other extremely reluctant site is highly unlikely. The on-site options are in-tank solidification as at Hanford or emplacement in bedrock caverns. Due to the less favorable climatic conditions, it is possible that the caverns will be excavated in the bedrock below the site. The caverns can be lined with concrete and suitable containers (which could be the cavern itself) may be filled by pumping in a concrete supergrout containing the waste. By forming a cap of concrete containing no waste, one creates in effect large sealed concrete monoliths with solid cement-encapsulated waste inside. Alternatively, concrete or metal canisters filled with waste can be simply loaded into the mined caverns. The waste form could be a concrete or related low temperature hydroxylated waste form or it could be a dilute glass. The latter is only likely if the country is somehow stampeded into the decision to build a "glass plant."

C. NFS wastes are solidified into low-temperature hydroxylated waste forms and shipped to Savannah River for storage and later disposed.

The possibility of disposal on-site at West Valley, NY, by grouting has been rated as quite feasible by the United States Geological Survey. Such an option, however, is unlikely to be examined unless a great deal of the completely artificial "urgency" to come up with a "solution" disappears fast. With a public fully informed about the options over a five-year period, and with financial incentives provided to the State and County (instead of being spent on unnecessarily sophisticated technology) the grouting option using an Oak Ridge plant could most quickly and cost-effectively achieve the disposal at the Nuclear Fuel Sciences (NFS) plant.

The other more likely alternative is that the liquids will be converted into a low-temperature hydroxylated form including concrete, by reacting inside a corrosion-resistant alloy canister. Again, sufficient political manipulation may lead to a glass waste form which could likewise be shipped to a federal repository, probably at Savannah River.

## Commercial Waste Reprocessing Plant and Repository

Since the U.S. has no commercial waste reprocessing site, nor is any likely till about 2000 A.D., there should be plenty of time to co-locate the reprocessing and disposal facilities. Using the same arguments as above, only one or both of these two sites appear likely: at the Nevada Test (Weapons) Site facility and the Savannah River site. The waste package design could literally be any of the many combinations discussed above, since the differences between different packages will be relatively small.

## Common Strategy Which Maximizes Safety and Minimizes Cost

Given the situation that exists in the U.S., the strategy--adopted by many other countries--which offers the maximum safety is:

- Store spent fuel (or reprocessed waste) for 50 years.

- Transport once only, reprocess, solidify and dispose on the same site.

- Increase steady public education (via regular channels of education) regarding hazards from waste and the mitigation strategies available.

- Continue modest *long-term research* effort to discover radically more cost-effective technologies, and execute subsequent development work to design and test such new "waste packages."

# REFERENCES FOR CHAPTERS 9 AND 10

1. National Research Council, Committee on Radioactive Waste Management, Panel on Waste Solification (1978), Solidification of High Level Radioactive Wastes, National Academy of Sciences, Washington, DC.
2. A. Jacobson and R. Pusch, "Deposition of High-Level Radioactive Waste Products in Boreholes with Buffer Substance.", KBS Technisk Rapport 03, Stockholm (1977).
3. G.J. McCarthy and M.T. Davidson, Ceramic Nuclear Waste Forms: I, Crystal Chemistry and Phase Formation, Bull. Amer. Ceram. Soc. 54, 782 (1975).
4. J.W. Braithwaite et al. (1979), Corrosion Considerations for Nuclear Waste Isolation Canisters, Scientific Basis for Nuclear Waste Management, Vol. 1, Ed. G.J. McCarthy, Plenum Press, NY, p. 283.
5. J.D. Sundquist (1979), Preliminary Corrosion Test of a Glass-Ceramic Candidate for a Nuclear Waste Canister, Scientific Basis for Nuclear Waste Management, Vol. 1, Ed. G.J. McCarthy, Plenum Press, NY, p. 289.
6. E. Mattson (1979), Corrosion Resistance of Canisters for Final Disposal of Spent Nuclear Fuel, Scientific Basis for Nuclear Waste Management, Vol. 1, Ed. G.J. McCarthy, Plenum Press, NY, p. 271.
7. J.W. Braithwaite et al. (1980), Corrosion Resistant Metallic Canisters for Nuclear Waste Isolation, Scientific Basis for Nuclear Waste Management, Vol. 2, Ed. C. Northrup, Plenum Press, NY, p. 377H.
7a. W.N. Rankin (1980), Canister Compatibility with Carlsbad Salt. Scientific Basis for Nuclear Waste Management, Vol. 2, Ed. C. Northrup, Plenum Press, NY, p. 395.
8a. R.E. Westerman (1981), Development of Structural Engineered Barriers for the Long Term Containment of Nuclear Waste, Scientific Basis for Nuclear Waste Management, Vol. 3, Ed. J.G. Moore, Plenum Press, NY, p. 515.
8b. F. Casteels et al. (1980), Scientific Basis for Nuclear Waste Management, Vol. 2, Ed. C. Northrup, Plenum Press, NY, p. 385.
8c. S.G. Pitman, B. Griggs and R.D. Elmore, "Evaluation of Metallic Materials for Use in Engineered Barrier Systems," Scientific Basis for Nuclear Waste Management, Vol. 3, Ed. J.G. Moore, Plenum Press, NY, p. 523.
9. National Research Council, Committee on Radioactive Waste Management (1980), A Review of the Swedish KBS-II Plan for Disposal of Spent Nuclear Fuel, NAS-NAE-NRC, National Academy of Sciences, Washington, DC.
10. Engineered Waste Package: Conceptual Design. Westinghouse Electric Corporation. Report to ONWI Contract E512-06400, March 1981.
11a. R.Roy. Presentation to ERDA team at University Park, PA, June 20, 1973. See Mendel, McElroy, Platt, "High-Level Radioactive Management," pp. 106-109, M.H. Campbell, editor, ACS, 1976.

11b. H.W. Nesbitt et al. (1981). "The Stability of Perovskite and Sphene in the Presence of Backfill and Repository Materials," _Scientific Basis for Nuclear Waste Management_, Vol. 3, Ed. J.G. Moore, Plenum Press, NY, p. 553.

12a. B. Allard et al. (1977),"Sortpion of Long-Lived Radionuclides in Clay and Rock," KBS Tech. Report, Stockholm, p. 55.

12b. B. Allard et al. (1980), "Possible Retention of Iodine in the Ground," _Scientific Basis for Nuclear Waste Management_, Vol. 2, Ed. C. Northrup, Plenum Press, NY, p. 673.

13a. E.J. Nowak (1980), The Backfill as an Engineered Barrier for Nuclear Waste Management, _Scientific Basis for Nuclear Waste Management_, Vol. 2, Ed. C. Northrup, Plenum Press, NY, p. 403.

13b. E.J. Nowak (1981), Composite Backfill Materials for Radioactive Waste Isolation by Deep Burial in Salt, _Scientific Basis for Nuclear Waste Management_, Vol. 3, Ed. C. Northrup, Plenum Press, NY, p. 545.

14. G.W. Bird and V.J. Lopata (1980), "Solution Interactions of Nuclear Waste Anions with Selected Geological Materials," _Scientific Basis for Nuclear Waste Management_, Vol. 2, Ed. C. Northrup, Plenum Press, NY, p. 419.

15. D.G. Brookins (1980), Clay Minerals Suitable for Overpack in Waste Repositories, _Scientific Basis for Nuclear Waste Management_, Vol. 2, Ed. C. Northrup, Plenum Press, NY, p. 427.

16. S. Komarneni and R. Roy (1980), "Superoverpack: Tailor-Made Mixtures of Zeolites and Clays," Scientific Basis for Nuclear Waste Management, Vol. 2, C.J. Northrup, Jr. (Ed.) pp. 411-418.

16a. A.E. Ringwood (1978), _Safe Disposal of High-Level Nuclear Reactor Wastes: A New Strategy_, A.N.U. Press, Canberra, Australia.

17a. S. Komarneni and D.M. Roy, "Effect of Layer Charge and Heat Treatment on Cs Fixation by Layer Silicate Minerals," J. Inorg. Nucl. Chem. $\underline{40}$, 893-896 (1978).

17b. S. Komarneni and D.M. Roy,"Shale as a Radioactive Waste Repository: The Importance of Vermiculite," J. Inorg. Nucl. Chem. $\underline{41}$, 1793-1796 (1979).

17c. S. Komarneni and D.M. Roy, "Hydrothermal Effects on Cesium Sorption and Fixation by Clay Materials and Shales," Clays and Clay Minerals $\underline{28}$, 142-148 (1980).

18a. S. Komarneni and R. Roy (1981), Hydrothermal Transformations in Candidate Overpack Materials and Their Effects on Cs and Sr Sorption, Nucl. Tech. $\underline{54}$, 118.

18b. S. Komarneni and W.B. White (1981) Hydrothermal Reactions of Clay Minerals and Shales with Cesium Phases from Spent Fuel Elements," Clays Clay Miner. $\underline{29}$ (in press).

18c. S. Komarneni and R. Roy (1981), Interactions of Backfill Materials with Cesium in a Bittern Brine Under Repository Conditions, Nucl. Tech. (in press).

19. R. Roy and S. Komarneni (1981), Design of Overpack (= Backfill) Materials, "NBS Workshop on the Research and Development Needs in Backfill for Underground Nuclear Waste Management," National Bureau of Standards Proceedings, Washington, DC.

20. C.F. von Weizsacker, "The Politics of Peril," Seabury, NY, 1978, Chapter 6.

21. H.C. Burkholder (1980), Waste Isolation Performance Assessment: A Status Report, Scientific Basis for Nuclear Waste Management, Vol. 2, Ed. C. Northrup, Plenum Press, NY, p. 689.

22. S. Rymochowicz (1977), A Collection of Results and Methods on the Leachability of Solidified High Level Radioactive Waste Forms, Hahn-Meitner Institute Report, HMI-B241.

23. J.E. Mendel et al. (1977), Annual Report on the Characteristics of High-Level Waste Glasses, BNWL-2252, Battelle, Pacific Northwest Laboratories, 99 pp.

24. C. Houser, I.S.T. Tsong, and W.B. White (1979), Characterization of Leached Surface Layers on Simulated High-Level Waste Glasses by Sputter-Induced Optical Emission, Scientific Basis for Nuclear Waste Management, Vol. 1., Ed. G.J. McCarthy, Plenum Press, NY, p. 131.

25. A.B. Harker, C.M. Jantzen, P.E.D. Morgan, and D.R. Clarke (1980), Tailored Ceramic Nuclear Waste Forms: Preparation and Characterization, Scientific Basis for Nuclear Waste Management, Vol. 2, Ed. C. Northrup, Plenum Press, NY, p. 139.

26. A.E. Ringwood, K.D. Reeve, and J.D. Tewhey (1980), Recent Progress on Synroc Development, Scientific Basis for Nuclear Waste Management, Vol. 2, Ed. C. Northrup, Plenum Press, NY, p. 147 (see also p. 165).

27a. G.J. McCarthy (1977), High-Level Waste Ceramics: Materials Considerations, Process Simulation and Product Characterization, Nucl. Technol. 32, 92.

27b. T. Adl and E.R. Vance, $CsAlSi_5O_{12}$ Possible Host for $^{137}Cs$ Immobilization, J. Mat. Sci. (in press).

27c. W.B. White et al. (1981), "Dissolution Rates of Monazite as a Function of pH,p,t." In preparation.

28. G.J. McCarthy et al. (1978), Interactions Between Nuclear Waste and Surrounding Rock, Nature 273, 216.

29. National Waste Terminal Storage Program Conference on Waste Rock Interaction, D.M. Roy, Chairman, Y/OWI/77/14268, July 1977.

30. R. Roy and O.F. Tuttle (1956), Investigations Under Hydrothermal Conditions, Physics and Chemistry of the Earth 1, Pergamon Press, p. 138.

31. B.E. Scheetz, W.B. White and S.D. Atkinson, Dissolution of Aluminum, Titanium and Zirconium-Based Crystalline Waste Form Components (Accepted for publication in J. Nuclear Technology).

32. J. Apps and N.G.W. Cook (1981), Backfill Barriers: The Use of Engineered Barriers Based on Geologic Materials, Scientific Basis for Nuclear Waste Management, Vol. 3, Ed. C. Northrup, Plenum Press, NY, p. 291.

33. H.W. Nesbitt et al. (1981), The Stability of Perovskite and Sphene in the Presence of Backfill and Repository Materials: A General Approach, Scientific Basis for Nuclear Waste Management, Vol. 3, Ed. C. Northrup, Plenum Press, NY, p. 131.

34. W.P. Freeborn et al. (1980), Shale Rocks as Nuclear Waste Repositories: Hydrothermal Reactions with Glass, Ceramic and Spent Fuel Waste Forms, Scientific Basis for Nuclear Waste Management, Vol. 2, Ed. C. Northrup, Plenum Press, NY, p. 499.

35a. S. Komarneni and B.E. Scheetz (1981), Hydrothermal Interactions of Basalts with Cs and Sr of Spent Fuel Elements: Implications to Basalt as a Nuclear Waste Repository, J. Inorg. Nucl. Chem. 43, 1967.

35b. B.E. Scheetz et al. (1980), Hydrothermal Interaction of Simulated Nuclear Waste Glass in the Presence of Basalts, <u>Scientific Basis for Nuclear Waste Management</u>, Vol. 2, C.J. Northrup, ed., Plenum Press, NY, p. 207.

35c. W.B. White et al. (1979), Brine-Waste Form Interactions Under Mild Hydrothermal Conditions, <u>Scientific Basis for Nuclear Waste Management</u>, Vol. 1, G.J. McCarthy, ed., Plenum Press, NY.

36. N. Sasaki, S. Komarneni, B.E. Scheetz and R. Roy, Backfill (=Overpack)-Waste Interactions in Repository Simulating Tests (Accepted for publication), <u>Scientific Basis for Nuclear Waste Management</u>, Vol. 4.

APPENDIX I.


by

John E. Mendel

The Battelle Pacific Northwest Laboratory


(Reproduced by permission from BPNL Report # 2764)

Appendix I.   GLASS AS A FULLY ENGINEERED WASTE FORM

## Glasses Made to Date

In Chapter 4, glass as a radiophase was treated in comparison with
other candidates.  Due to the enormous amount of development work on glass,
however, this development has produced many large "ingots" of glass all
over the world.  The composition and processes used have been described in
Chapter 4.  We turn now to the products and their properties.  Fig. 1 shows
a lab-scale radioactive sample of borosilicate glass which has been exposed
to a very high induced radiation dose.  Figs. 2a and 2b show engineering
scale examples of such non-radioactive and radioactive, respectively, glass
blocks cast into stainless steel canisters.

## Physical and Mechanical Properties of Waste Glass

The physical and mechanical properties of radioactive waste glass are
innate properties which may be treated generically.  They are relatively
independent of composition and are of much the same magnitude for most
waste glasses.

## Viscosity and Softening Point

Waste glasses are formulated to have a viscosity of 5 to 200 poises at
the processing (melting) temperature, which is usually within the range of
950 to 1150°C.  The viscosity of waste glasses is contrasted with that of
commercial glasses in Fig. 3.

Some important properties of waste glass can be related to viscosity.
The softening temperature (Littleton softening temperature) of glasses is
defined as the temperature at which the viscosity is $10^{7.6}$ poises.  This is
the temperature above which the glass cannot support its own weight and
begins to slump.  Cracks in the glass will heal rapidly at the softening
temperature, which is about 575 to 650°C for most waste glasses.  The glass
transition temperature is the temperature at which the viscosity is about
$10^{13}$ poises.  For waste glasses this usually is in the range of 500 to
550°C.  As described earlier, glass is plastic above the glass transition
temperature and thus is not susceptible to fracture.

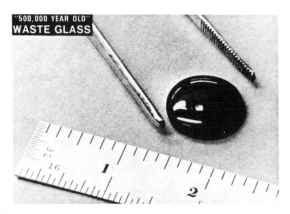

FIGURE 1.  Waste Glass Specimens Doped with 8 wt% $^{244}Cm_2O_3$ After Cumulative Radiation Dose Corresponding to 500,000 Years Storage

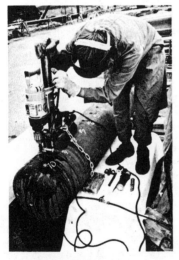

a.  CORE-DRILLING TO OBTAIN WASTE GLASS SAMPLES FOR CHARACTERIZATION FROM CANISTERS PRODUCED IN NONRADIOACTIVE PILOT PLANT

b.  VIEW THROUGH LEAD GLASS WINDOW OF 8-YEAR OLD CANSITER OF BOROSILICATE WASTE GLASS WHICH CONTAINED 1,400,000 Ci WHEN PREPARED. (NOTE PLUGGED CORE-DRILL HOLES FROM WHICH SAMPLES OF RADIOACTIVE GLASS WERE REMOVED FOR CHARACTERIZATION.)

FIGURE 2.  Waste Glass Canisters Prepared in Nonradioactive and Radioactive Pilot Plants at PNL

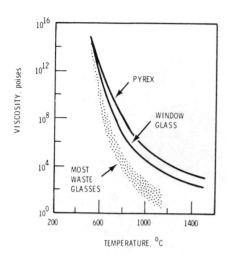

**FIGURE 3.** Viscosity of Typical Waste Glass

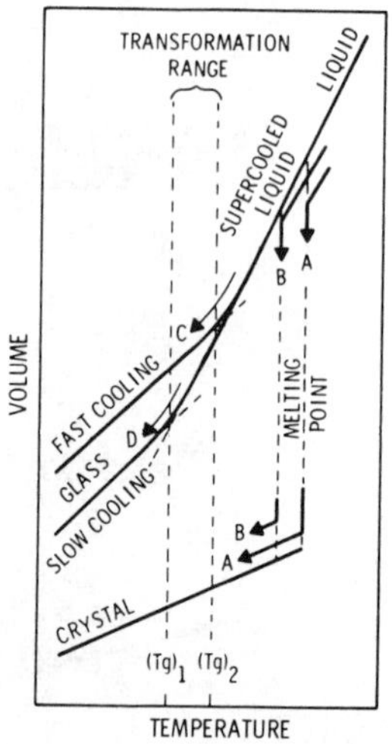

**FIGURE 4.** Volume-Temperature Relationships for Glasses, Liquids, Super-Cooled Liquids, and Crystals

## Density

Density is probably the only physical and mechanical property of waste glass that will be routinely measured at an operating plant. The bulk density, as determined from the weight and height of the glass column in the canister, can serve as a quality control check.

The density of glass is an additive function of the constituents (3). Within limits the density of a glass can be predicted from empirically-derived density factors for each constituent. The density factors are related to atomic weight. Thus, waste glasses containing high loadings of fission products, which have higher atomic weights than most usual glass constituents, will be more dense than those with low loadings of fission products. Certain special glass constituents, such as zinc or lead, also markedly increase density.

Common commercial soda-lime-silica glass has a density of about 2.5 g/cc. Depending on waste composition and waste loading, radioactive waste glass densities are usually in the range of 2.5 to 3.3 g/cc. Zinc borosilicate waste glasses may have a density of 3.7 g/cc or higher and the density of a lead borosilicate waste glass may exceed 4.8 g/cc. Radiation over long periods of time can cause changes in waste glass density; maximum change equals approximately 1% (see section on Density Change).

For a given glass composition, the density of the final product may vary over 0.1 g/cc depending on the time-temperature curve of its cooling through the transformation range (Fig. 4). Thus, bulk density cannot serve as a precise quality control check on glass composition. It can, however, be a convenient quality control check for gross porosity or voids in the canister.

## Friability

Friability refers to the fracture behavior of waste glass. Fractures increase the surface area, and the amount of activity released in a leaching situation is proportional to the exposed surface area. Also, the fractures may produce a small fraction of less than 10 μm diameter particles, which are respirable. Friability becomes a factor in safety analyses only when the canister is breached. Canister breaching requires either very high accidental impacts, which are improbable, or; corrosion through the canister wall, which should not occur until well after the waste glass is in a geological repository.

Friability of waste glass is usually measured by impact tests of various kinds, in which either weights are dropped on a glass specimen, or the glass specimen itself is dropped onto an unyielding surface. The most comprehensive impact tests on waste glass yet performed are described by Smith and Ross (4). In these tests, the glass was cast in stainless steel canisters either 2 in. in diameter by 4 in. long or 6.6 in. in diameter by 52 in. long. The large canisters were dropped onto a concrete pad from heights up to 30 feet (maximum impact velocity equalled 44 ft/sec). The smaller canisters were released in the path of a 75-1b. granite block mounted on a rotating arm. Impact velocities up to 117 ft/sec (80 mph) were achieved in this apparatus. None of the large canisters failed in the impact tests; some of the small canisters had small cracks after impacts at 76 and 117 ft/sec, but weight loss measurements showed that no glass escaped through the cracks. After the tests each of the canisters was opened and the particle size of the contained glass was determined. Figure 5 shows the amount of sub 10 μm particles formed as a function of impact velocity. It should be noted that both vitreous and devitrified specimens of the same glass composition were included in these tests and no difference in fracture behavior was observed. Therefore, the curve shown in Fig. 5 is based on data from both vitreous and devitrified glass.

Friability is also a function of internal stresses. These are difficult to define but include residual thermally-induced stresses and, over the long term, radiation-induced stresses. As mentioned in the preceding section, radiation can cause slow changes in density. In devitrified glass or crystalline materials, particularly, differing radiation-induced density changes in adjoining phases could lead to localized stresses. The subject of radiation-induced density change is covered in more detail in the section on Density Change.

Friability is relatively independent of glass composition. Impact tests comparing waste glasses and commercial soda-lime-silica glasses yielded similar results (5).

## Thermal Conductivity

Thermal conductivity is another physical property of waste glass that is relatively independent of glass composition. The thermal conductivity of most waste glasses falls within the limits shown in Fig. 6. The thermal conductivity increases gradually with temperature until the softening point of the glasses is reached. Above the softening point, the increase in the thermal conductivity as a function of temperature is more rapid.

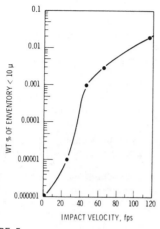

**FIGURE 5.** Effect of Impact Velocity on Fraction of Respirable Particles Formed

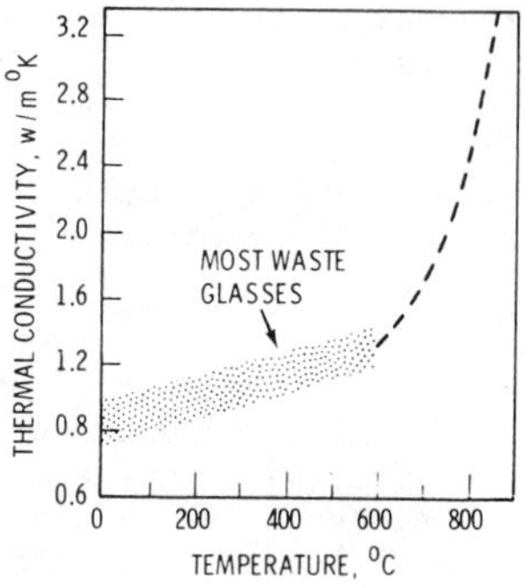

**FIGURE 6.** Thermal Conductivity of Waste Glasses

The importance of thermal conductivity is that it determines the temperature profile between the glass at the center of the canister and the glass at the canister wall. Higher thermal conductivities give less steep temperature profiles.

## Thermal Expansion Coefficient

The thermal expansion coefficient of waste glasses is similar to that of commercial soda-lime-silica glasses. Representative thermal expansion coefficients of waste glass are shown in Table 1. The thermal expansion coefficients are relatively independent of composition. Sodium probably has the greatest effect on thermal expansion of any of the waste constituents. Decreasing sodium content usually decreases the thermal expansion coefficient. (Commercial low expansion glasses, such as Pyrex, have a very low sodium content.)

### TABLE 1

#### Typical Thermal Expansion Coefficients of Waste Glass

| Waste Glass Type | Thermal Expansion Coefficient (25-400°C)   $\alpha \times 10^{-7}/°C$ |
|---|---|
| Borosilicate | 80-100* |
| Glass-ceramic | 100-120** |
| Phosphate | 90-110*** |

For comparison, the values for soda-lime-silica bottle glass and Pyrex are $90 \times 10^{-7}$ and $30 \times 10^{-7}$, respectively.

*Ross (6-7).    **De (8).    ***Thompson (9).

A low thermal expansion coefficient gives increased thermal shock resistance and decreases the time required to anneal a block of glass, and thus would appear to be a desirable characteristic. On the other hand, a low thermal expansion coefficient will further increase the mismatch in thermal expansion coefficient between the waste glass and its canister, and thus will increase the stress in the canister wall. The thermal expansion coefficients of waste glass, as shown in Table 1, represent a good compromise between the opposing requirements. In any event, effecting significant changes in the thermal expansion coefficient would probably be very difficult. Sodium, for instance, is almost always present in waste glasses in relatively high concentrations, either because it is an unavoidable constituent of the waste or because it is needed for viscosity control.

## Thermal Effects and Devitrification

The facilities for waste vitrification and handling of the canisters of waste glass are designed to avoid deleterious temperatures. Thus design limits, based on known glass and canister properties, are placed on the maximum temperatures allowed during waste glass manufacture and storage. For accident analyses, it is necessary to know what will occur when the design temperatures are exceeded.

### Volatility

After manufacture, volatility from waste glass is a factor only in accident analyses. Waste glass manufacturing temperatures are several hundred degrees above the maximum design storage temperatures, which assures that any volatiles that might pressurize the canisters during storage have been removed. However, accidents can be postulated, involving external heat sources or clusters of uncooled waste canisters, in which the temperature rises far above normal storage or handling temperatures. Volatility data are available for analysis of these postulated accidents. An example of such data is shown in Fig. 7. The data demonstrate that volatility is very temperature dependent. Analysis of postulated accident scenarios is continuing, but accidents in which waste glass could be exposed to temperatures as high as those shown in Fig. 7 will be extremely rare.

Volatility experiments with waste glasses show that cesium is the most volatile radioisotope. Although ruthenium often exhibits significant volatility in radioactive processes, its volatility from waste glasses is less than that of cesium.

### Devitrification and Phase Separation

Major thermal effects on waste glass are devitrification and phase separation. Both phenomena are very dependent on glass composition, but some effects are common to most compositions.

Devitrification. Devitrification is the formation of crystals caused by the rearrangement of certain atoms in the glassy matrix. Crystals form because their ordered structure has a lower free energy. The ordered

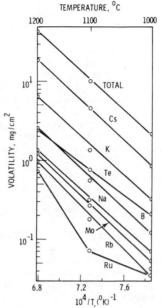

**FIGURE 7.** Volatility of Waste Consti.
(Gray, 1976)

structure is more thermodynamically stable than the random network of the
glass. The amount and identity of the crystals formed depends on the
initial composition of the glassy matrix. The rate of crystal formation is
time and temperature dependent. Measurable devitrification of waste
glasses generally occurs only from 950 to 500°C. Above 950°C, the crystals
redissolve.* Below 500°C, the viscosity of the glassy matrix is so high
that crystallization is diffusion-limited to extremely low rate. The
maximum rate of formation of most crystalline species occurs somewhere
between 800 and 850°C but the equilibrium yield of crystals is low. At
lower temperatures the rate of formation decreases but the equilibrium
yield of crystals increases. It is possible to define approximate equilibrium yields at 700°C in experimentally practical times (1 year). At lower

---

*There are some exceptions to this statement. Fission product ruthenium and
palladium are insoluble in waste glass at all temperatures. The fission
product oxide $CeO_2$ does not completely dissolve in waste glass until a
temperature of about 1200°C is reached. Spinels, having the general
formula $A_2BO_4$, where A can be Fe or Cr and B can be Ni, Zn, Fe(IIO, etc.,
apparently may even exhibit a retrograde solubility above 950°C in certain
waste glass compositions. Regardless of their high temperature behavior,
the effects of these species are indistinguishable from the other devitrification species at the handling and storage temperature we are concerned
with here.

temperatures the experiments cannot be completed in practical times. But the temperature effect above 700°C exhibits Arrhenius behavior and can be extrapolated to lower temperatures. Extrapolation shows that the kinetics are extremely slow at geologic storage temperatures. Below approximately 200°C, measurable crystallization (anhydrous) would require millions of years (10). The existence of natural glasses which are millions of years old is an indication that the extrapolation is justified.

It is important to understand what devitrification does to waste glass. Figure 8 shows photomicrographs of a typical waste glass before and after devitrification. It is only on the microscopic scale that the effects are truly apparent. Even totally devitrified waste glass usually includes only 30 to 40 vol % crystals, as compared to 95 vol % in commercial

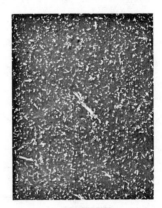

 VITREOUS                    DEVITRIFIED

**FIGURE 8.** Photomicrographs of Vitreous and Devitrified Waste Glass

Because the amount and identity of the crystalline species formed during devitrification are so dependent on composition, a large amount of work has been done to identify the various species that may form in specific waste glasses. A review of that work is beyond the scope of this report; however, some general observations may be made:

glass-ceramics. To the naked eye the devitrified waste glass is unchanged except that its luster is dulled and the color may change.

Changes in friability due to devitrification are slight, usually within experimental error. Devitrification should not be equated with disintegration. Properly formulated waste glass retains its coherence and strength when devitrified.

Devitrification usually increases the overall leach rate of waste glasses. The increase may occur in the residual glass phase (due to depletion of silica) or in one of the crystalline phases that is formed, like strontium molybdate (11). The increase in leaching of borosilicate waste glasses after devitrification is usually no more than a factor of 2 to 5, although in some instances an increase up to a factor of 10 has been observed in waste glass after maximum devitrification at 750°C as shown in Fig. 9. Figure 9 also demonstrates that there can be considerable variation in the leach rate of nondevitrified waste glass.

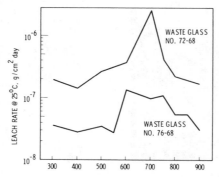

FIGURE 9. Leach Behavior as a Function of Devitrification for Two Representative Borosilicate Waste Glasses

Because the amount and identity of the crystalline species formed during devitrification are so dependent on composition, a large amount of work has been done to identify the various species that may form in specific waste glasses. A review of that work is beyond the scope of this report; however, some general observations may be made:

• The species formed on devitrification of borosilicate waste glasses are often crystalline silicates; boron always remains in the glassy matrix. The end result is a boron-rich residual glassy phase, which has a somewhat higher leach rate than the initial glass due to its higher boron content.

- A larger proportion of the high valence, high atomic weight cationic constituents tend to appear in the crystalline silicate species. Sodium, potassium, calcium, etc., are relatively absent from the crystalline silicate devitrification species. In general, however, other anions that may be present in the glass, such as molybdate, phosphate, or fluoride, exhibit a strong tendency to form crystalline devitrification species which, in contrast to the silicates, usually preferentially contain alkalis and alkaline earths. The identities of the exact crystalline compounds that will form from a given glass composition cannot yet be predicted with confidence; they must be determined empirically.

- Secondary devitrification reactions have also been observed in which a rapidly-formed crystalline species is slowly converted into another crystalline species.

- At optimum devitrification conditions, crystals may form several hundred micrometers in the longest dimension. Micro-cracking in the surrounding glass matrix may occur, due to mismatching thermal expansion coefficients.

It is apparent that devitrification of waste glasses is a complex phenomenon. Devitrification can be prevented by cooling the glass quickly and subsequently maintaining the glass at temperatures below approximately 500°C. However, the primary concern is: Need devitrification of waste glass be avoided? And the answer seems to be no, based on all available data. The only significant devitrification effect is on leach rate, and the magnitude of this effect can be defined for any given glass with a few simple tests. To date, these tests have always shown no more than a factor of 10 increase in the leach rate of cesium, the radioisotope whose leach rate is most sensitive to devitrification effects. It should be emphasized that if the decision is made to "accept devitrification," only a limited central region (perhaps 20 vol %) of the waste glass in the canister would be affected. The waste glass at the canister wall would remain vitreous because the maximum allowable wall temperature during handling is only 375°C.

A large amount of work is continuing on the devitrification of waste glass. Two areas in particular are receiving attention. More needs to be learned about the location of the actinides in devitrified waste glass. The effects of the nonuniform radiation pattern caused by the possible concentration of some actinides in certain crystalline species are now being measured. Hydrothermal alteration, as opposed to thermal devitrification discussed so far, is also being studied. Based on analogies with naturally-occurring glasses, the possible enhancement of crystallization by water at geologic storage temperatures is being examined. This topic is discussed more fully in the section entitled Effect of Temperature and Pressure.

Phase Separation. Phase separation is another time-temperature dependent phenomenon involving atomic rearrangement. It is a liquid-liquid phase separation. Borosilicate glasses are particularly susceptible to phase separation, forming a boron-rich and a silica-rich phase (12). There is evidence that some borosilicate waste glasses are phase separated as formed.

FIGURE 10. Transmission Electron Micro-
graph of Phase-Separated Waste Glass

Phase separation is difficult to study because the dimensions of the separated phases are usually less than 500A. Figure 10 is a photomicrograph of a typical waste glass illustrating probable phase separation. Two types of phase separation are known in borosilicate glasses. In one the borate-rich phase is distributed as discrete disconnected globules in the silica-rich phase. In the other, the borate-rich globules are connected to form a continuous network. It may be postulated that the former type of phase separation occurs in waste glasses, since if the borate-rich phase was continuous, higher leach rates would be expected than those actually observed.

Molybdenum and sulfate also exhibit phase separation in borosilicate waste glasses. A molybdenum-rich salt phase that also contains sodium, cesium, and strontium separates. As is typical of phase separation phenomena, the molybdenum-rich phase is completely miscible in waste glasses above about 1150°C. At lower temperatures the phase separation is composition dependent. If the concentration of molybdenum and sodium is high enough in borosilicate waste glasses, the molybdenum-rich phase will separate out on a macroscale and float to the surface of the molten glass. As noted on page 17, reducing conditions can be used to avoid the separation of the molybdenum salt phase.

As in the case of molybdenum, sulfate tends to separate as a sodium salt phase, which may also contain other alkali metals, alkaline earths, and chromates. Immiscibility of $Na_2SO_4$ becomes apparent when the $SO_3$ content of borosilicate waste glass exceeds 1.2% (13). If the sulfate content is expected to be higher than this, the presence of soluble $Na_2SO_4$ in the product glass can be prevented by adding reducing agents, such as C, to the melt. Sulfate will then be reduced to gaseous $SO_2$ at glass melt temperatures. Lead-borosilicate glass compositions have also been described for the immobilization of sulfate-containing radioactive waste (14).

Phosphate glasses are less susceptible to phase separation. In particular, molybdenum does not separate out in phosphate glasses. The French have developed phospho-silicate glass compositions for the incorporation of very high molybdenum-content radioactive wastes (15).

## Leaching

The known chemical durability of the ancient manmade glasses of the Mesopotanians, Egyptians, Greeks, and Romans for 2000 to 4000 years, even during immersion in seawater for over 2000 years in the case of Greek and Roman glasses, was one of the original incentives for developing waste glasses. Following this lead, measurements of leach rate have been key factors in the selection of waste glass compositions.

Intuitively, low leach rates are desirable. Two questions arise, however, in quantifying what is apparent intuitively: (1) How low must the leach rates be? and (2) Can leach rates measured in the laboratory be interpreted and extrapolated to predict long-term behavior in actual geologic situations? The first question is quite complex and is the subject of on-going risk analyses which are outside the scope of this report. Answering the second question, however, is precisely the ultimate goal of the leach tests that are being made on waste glasses. The answer is not yet complete. The present status of the investigations will be described in the following sections.

Table 2 shows the factors that can significantly effect the value obtained in a given leach rate. Some of these factors are intrinsic, related only to the waste glass itself and its condition. Others are extrinsic; they are external factors which can vary independently of the waste glass. The factors in Table 2 are interrelated. Understanding the leaching of waste glass rests on knowing the type of influence each factor has in relation to the others.

## TABLE 2

### Factors Affecting Leach Rate Measurements

| Intrinsic | Extrinsic |
|---|---|
| Glass Composition | Temperature |
| Thermal History | Pressure |
| Radiation Effects | Composition of Leachant |
| Physical Form | (including pH and Eh) |
| Surface Character | Flow Rate of Leachant |

## Mechanisms

The basic mechanisms of glass leaching are well known. Many commercial glass studies have clarified these mechanisms since chemical durability (resistance to leaching) is a fundamental requirement for many commercial glass applications. Although the basic mechanisms are known, understanding of the quantitative effects of the various factors as applied to specific situations is still lacking in many cases.

The primary leaching mechanism below about 80°C is a diffusion-controlled ion exchange in which hydrogen or hydronium ions exchange for cations in the glass lattice. The overall result of this reaction is a depletion of cations from the glass surface leaving a hydrous silica layer, which also contains some slow-diffusing multivalent cations. As this silica-rich surface film develops, it functions as an increasingly protective barrier which progressively slows down the rate of leaching. Finally, the rate of leaching is slowed until a second mechanism assumes the rate-controlling function. The second mechanism is surface corrosion. Surface corrosion refers to erosion of the silica-rich film by a combination of chemical dissolution (the solubility of silica in deionized water is 18 ppm) and mechanical sloughing-off of hydrous increments of the film. At room temperature, surface corrosion results in a much slower reaction than diffusion, but as the temperature increases, surface corrosion apparently increases at a faster rate than diffusion. At temperatures above about 80°C surface corrosion is usually the controlling mechanism (16).

The following expression is often used to express the leach rate of silicate glasses (17):

$$Q = at^{\frac{1}{2}} + bt$$

Where: $Q$ = the quantity of a specific cation that has been removed from the glass; $t$ = the time glass has been in contact with leachant; $a,b$ = constants. The first term represents the contribution due to diffusion and the second term, that due to surface corrosion.

The foregoing explanation of the mechanisms involved in leaching is based mainly on studies of simple alkali silicate glasses which have leach rates that are high enough to be measured conveniently. Commercial glasses contain constituents added specifically to increase leach resistance, such as alkaline earth and aluminum oxides. These additvives increase the chemical durability so much that it is very time-consuming to make measurements to corroborate the leaching mechanisms. Usually accelerating techniques, such as elevated temperatures or high acidity, are resorted to, which complicate interpretation in terms of fundamental mechanisms. Overall, the results indicate that the basic mechanisms inferred from the simple glasses are useful guidelines for interpretation of the data, especially for extrapolation to longer time behavior, but other factors are also involved. A principal factor is composition of the leachant. In actual leaching situations the water in contact with waste glass will be impure because: (1) natural waters contain dissolved ions, and (2) unless the water is moving very rapidly, dissolved ions from the glass will concentrate at the surface. These ions can participate in complex back reactions which usually, but not always, slow the leach rate of the glass (18).

The ultimate conclusion is:  In order to evaluate a glass for a specific use, the test program should duplicate as closely as possible the conditions of that use (19).  This is even more true for radioactive waste glasses than for commercial glasses, because of the complexity of the waste glass compositions and their applications.

## Techniques for Waste Glass Leach Rate Measurements

The majority of leach rates that have been measured on waste glasses to date have used deionized water that is replenished on a preset schedule. The leach-rate data obtained in this manner serve as an index of relative leach resistance for the comparison of waste glasses.  These data can be used to estimate the behavior of waste glass in certain situations, such as:  (1) canister failure in a water storage basin, (2) a transportation accident in which the failed canister lands in a river, and (3) even as an upper limit in some geologic repository situations.  Leach rate data more pertinent to waste glass behavior in geologic repositories require a closer approximation of repository conditions and are just now beginning to be reported.

Leach tests can be divided into two principal types, dynamic and static.

Dynamic Leach Tests.  In dynamic leaching the test specimen is either exposed to flowing leachant, or to aliquots of leachant which are re-plenished periodically.  The goal is to minimize interfering surface reactions; relatively constant dilute conditions are maintained in the leachant contacting the specimen.  Dynamic tests furnish data for inter-comparison of waste glasses and for evaluation of accident behavior in flowing water.  The standard leach test proposed by the IAEA (20) and the Soxhlet test (21) are both dynamic tests.

The *proposed IAEA test* uses waste glass specimens immersed in water at room temperature.  The water is removed, analyzed, and replaced with fresh water at set intervals.  The intervals between leachant changes are one day initially, but are gradually increased as the test continues.*

In the *Soxhlet test* the waste glass specimen is contacted with flowing distilled water at a temperature that depends on design of the apparatus, but is usually in the range of 80 to 100°C.  Because of the elevated temperatures employed, test results are obtained quickly; duration of the tests is usually 72 hours or less.

Static Leach Tests.  In static leach tests the test specimen is exposed to stagnant water.  Static leach tests are of most value in evalua-ting the behavior of waste glass in the event of water intrusion into a waste repository, where the water flow rate would usually be very low.  The high temperature-high pressure tests being made with waste glasses are usually static tests.  The goal is to determine the final equilibrium state of the system.

---

*Users of the proposed IAEA test find that the leach rate they measure is somewhat dependent on the frequency at which the water is changed (2).  The IAEA test is being modified by a working group of the International Stan-dards Organization (ISO) and it will probably be recommended that the fre-quency of water changes be kept uniform throughout the duration of the test.

Reporting Leach Tests Results. Leach tests usually measure either
(1) weight lost by the glass specimen as a function of time or (2) the
concentration in the leachant of one or more of the leached ions as a
function of time. If weight loss is measured, the results are reported
simply as percent weight loss or as a corrosion rate, penetration depth per
unit time. Such results are useful for comparison purposes, but are an
over-simplification since congruent corrosion of all waste glass constituents is assumed. In actuality, incongruent leaching always occurs, at
least in the initial stages of leaching. Some alkali metal and alkaline
earth ions may leach at rates 10 to 100 times faster than cations of higher
valence, such as the rare earths. The extent of incongruent leaching can
be determined if the concentration of different ions in the leachant is
determined. This is the procedure used to obtain fundamental data which
give information about leaching mechanisms. It is the procedure used in
the IAEA proposed standard leach test.

The leach rate, as calculated from analysis of the ions in the leachant
is usually expressed as follows:

$$L = \frac{a_n}{A_o} \frac{W}{St_n} = g^*/cm^2-day$$

where: a = amount of a specified isotope leached during the nth leachant
renewal period; A = amount of the specific isotope initially present in
test specimen; W = weight of test specimen, grams; S = geometrical surface
area of test specimen, cm; t = duration of leachant renewal period, days.
The leach rate can also be expressed as a penetration rate, cm/day, which
is obtained by dividing L by the density of the glass.

Figure 11 shows the typical range of leach rates obtained for waste
glasses when leached according to the IAEA proposed standard leach test (or
procedures that have basic resemblances to the IAEA test). An indication
is also given of the usually observed relationships in the leach behavior
of important radioisotopes. Leach rate is quite dependent on glass
composition, but the leach behavior of most waste glasses lies somewhere
within the range shown in Fig. 11. As Fig. 12 shows, the same leach data
can be plotted as cumulative penetration, which is sometimes thought to
give a more graphic interpretation of what the leach rates mean.

## Effect of Flow Rate and Potential for Back Reactions

Various situations can be hypothesized in which waste glass could come
in contact with water. The flow rate of water past the waste glass can
vary widely in the hypothesized situations, ranging from the rate of a
river running past particles of waste glass to zero flow rate in a
repository containing water that never moves. The short-term effect of

---

*The assumption being made is that all of the glass constituents leach at
the same rate as $a_n$. Of course they do not, thus different leach rates
can be obtained for the same glass sample, depending upon which specific
isotope is analyzed. The correct way to express a waste glass leach rate
is: "g of glass/cm$^2$-day, based on the behavior of $^{137}Cs$" (or $^{90}Sr$,
whatever isotope was analyzed in the leachant).

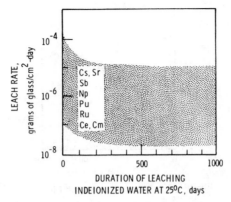

**FIGURE 11.** Typical Waste Glass Leach Rates

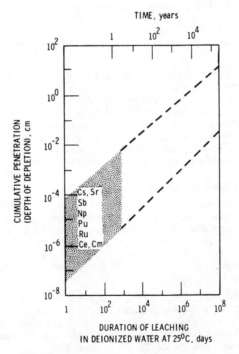

**FIGURE 12.** Typical Waste Glass Leach Rate
Data Expressed as Cumulative Penetration

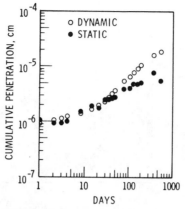

FIGURE 13. Effect of Leachant Flow on Leach Rate

flow rate is that the leach rate decreases as the flow rate decreases. This is because of "back reactions," including formation of a saturated hydrous film on the surface and even the possible formation of precipitates which protect the surface from further reaction. Such short-term effects are well known in waste glass leaching. Figure 13 shows an example of this effect.

At slower flow rates the pH of the solution in contact with the glass surface rises because hydrogen ions are used up in the leaching reaction. The rate of attack on glass increases significantly when the pH raises above 8.5 to 9.0 (22). An equilibrium pH is observed in commercial glass bottles holding static water that may be as high as 9.5 (23,24). Similar equilibrium pHs are observed with some waste glasses. Waste glass is being further tested, but present indications are that even though the pH may be quite high, steady-state stagnant systems exhibit very low leach rates.

## Effect of Temperature and Pressure

The effect of temperature on leach rate must be considered because the waste glasses may be self-heating to various degrees. The leach rates of commercial glasses increase with temperature, usually by a factor of 10 to 100 for each 100°C increase in temperature (18). Similar increases are observed for waste glasses although the rate of increase tends toward the lower end of the range, i.e., about a factor of 10 for each 100°C increase in temperature (25).

Leaching temperatures above 100°C could exist in the geologic repository, although it is by no means certain that they will. (While the waste glass is on the surface, the maximum leaching temperature is limited to 100°C by the boiling point of water.) Within a relatively wide latitude the geologic repository can be designed to operate at any temperature desired. Cooling will be by natural conduction after the repository is sealed, and the maximum temperature can be controlled in the following ways:

- The time of repository sealing and natural conductive cooling can be supplemented with convective or forced air cooling until the repository is sealed.

- The heat content of the individual canisters can be limited by a lengthened interim storage period before the waste glass is emplaced in geologic repositories, or by a reduction of the waste loading in canisters of high-heat wastes.

- The dimensions of the storage array can be adjusted. Cooler temperatures can be maintained by spreading out the waste. Storage efficiency can be maintained by interspersing low-heat producing wastes.

- The effective heat transfer can be increased between the waste glass capsule and the host rock. (For instance, wet, rather than dry, conditions will increase heat transfer and thus lower the temperature of the waste.)

The maximum temperature permitted in the repository will be determined through tradeoffs between economics and design or operational difficulties. Minimum cost would seem to be associated with high waste loading in small mine volumes, but several other factors have to be considered:*

- Heat effects must not adversely affect the structural integrity of the mine while portions of it are still being filled and/or occupied by workers.

- Thermal and radiation effects on repository host rocks must be considered, including dehydration, recrystallization, changes in thermal conductivity, and physical strength.

- Thermal expansion must not cause fracturing of either the repository rock, or impervious layers around the repository, which can adversely affect its integrity.

- Reaction rates of water with the host rock and with canisterized waste glass must be considered.

These factors are still being evaluated and the final design temperature of the repositories will not be known until the evaluations are complete. Reaction rates between waste glass and the host rock will be one input to the evaluation.

In autoclave tests, reactions which can be characterized as hydrothermal alteration have been observed with waste glasses above about 150°C. In these reactions, crystalline species are formed at the surface of the glass. The reactions are probably a combination of atomic rearrangement in the leached portions of the glass matrix plus deposition of precipitates formed from leached ions. Potential for the amount and number of latter reactions is greatly enhanced when the waste glass is in contact with a geologic medium, as discussed in the section on Effect of Leaching Media.

---

*Retrievability considerations will also limit temperature, but it is possible only a limited number of initial canisters will be designed to be retrievable.

Hydrothermal experiments involving waste glass and candidate geologic hosts, in particular salt, have been underway at PNL for about two years. Most of these tests have been conducted at 250 to 350°C to obtain steady-state conditions rapidly (one to three weeks) at the extreme high end of the potential temperature range. As stated above, the design temperatures for the repository have not been fixed; data obtained in the hydrothermal experiments, and at lower temperatures, will serve as one input factor to the decision on what the design temperature should be. A recent topical report by J.H. Westsik, Jr., and R.P. Turcotte describes the initial PNL hydrothermal tests (26). Data in the report demonstrate the following:

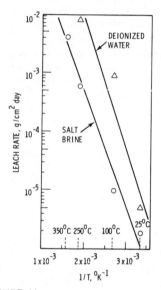

FIGURE 14. Effect of Temperature on HLW Glass Leach Rates in Salt Brine and Deionized Water

- The reaction rate of waste glass and aqueous solutions is a strong function of temperature. This is shown in Fig. 14.

- Other "stable" materials, including candidate geologic media, react at rates similar to waste glass in the hydrothermal tests, as Fig. 15 illustrates.

- The reaction mechanisms predominant in salt brine are different from those which predominate in pure water.

Figure 16, a and b, shows that reaction with the glass is slower in salt brine than in water perhaps because of decreased $H^+$ activity in salt brine. However, the ions which are released from the glass tend to stay in solution in salt brine because of chloride complexing, whereas in pure

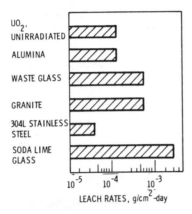

**FIGURE 15.** Comparative Leach Rates of
Several Materials in Salt Brine at 250°C

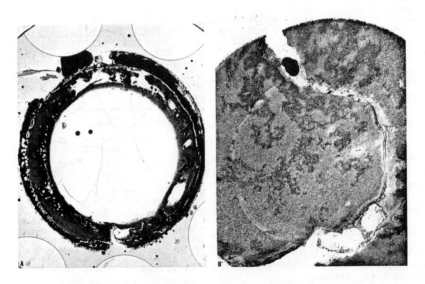

**FIGURE 16.** Cross Sections of Waste Glass After Exposure to
Salt Brine and Deionized Water for 1 wk at 350°C

water they tend to recombine to form new crystalline compounds. Identification of the new crystalline compounds is just beginning.

The preliminary conclusion is that until the hydrothermal reactions are understood in more detail, they should be minimized, since both the geologic host rocks and the waste glass can be significantly altered.* The reactions can be minimized if only small amounts of water are present while the repository is hot, or if the repository temperature never exceeds a certain predetermined limit, or if a multiple barrier system is used which prevents water from contacting the waste for several hundred years.

## Effect of Leaching Media

Leaching media that need to be considered in evaluating the leach resistance of waste glass fall into three major classifications, as shown in Table 3.

### TABLE 3

#### Potential Leaching Media for Waste Glasses

| Classification | Potential Method of Encountering Media | Possible Variations in Media |
|---|---|---|
| 1. Deionized water | Canister leak in water basin | Minor |
| 2. Nonsaline natural water | Canister failure during a transportation accident. Disposal in a nonsalt geologic repository* | Large major variables are pH, Eh, carbonate-bicarbonate, sulfate, calcium, and chloride (minor concentrations of chloride are ubiquitous in nature) |
| 3. Saline natural water | Disposal in a salt repository or in seabed* | Large major variables are pH, Eh, sodium, magnesium, and sulfate |

*Canister failure by corrosion is assumed to occur within a finite time.

Even though the variations possible in classifications 2 and 3 are very large, prohibiting experimental testing of all the possibilities, it is important to note that in the tests performed to date with various waste glasses, the leach rates measured in deionized water have always been equal to or higher than those measured in classification 2 or 3 media.** This is

*The discussion has disregarded the presence of the canister. This is a conservatism, since it is probable that canisters of present design will provide a significant barrier. (See comparative corrosion rates in Fig. 15.) If the canisters were designed especially for this service, different materials of construction would probably be used, such as titanium.

**An exception may be very low pH natural waters, known to occur, but which have not yet been tested as leachants for waste glass. The effect of pH on waste glass leaching is discussed later.

apparently due to back reactions of the type discussed in the section titled Effect of Flow Rate and Potential for Back Reactions. It may be concluded that the leach rates measured with the Soxhlet test and the proposed IAEA standard test, using deionized water, yield conservative results. They represent probable maximum leach rates. This conclusion is born out by the results of Canadian tests (27), which are the only experience to date with the *in situ* contact of waste glass with natural water.

In the Canadian tests, waste glass was buried approximately 10 to 12 feet underground in a swampy portion of the Chalk River nuclear reservation in 1960. The flow rate of groundwater past the buried waste glass waste was measured to be 200 ft/yr by a tritium innoculation technique. The groundwater downstream of the waste glass has been monitored for $^{90}$Sr since the burial. The leach rate calculated from the groundwater samples had decreased to $5 \times 10^{-11}$ g glass/cm$^2$-day by 1968 and has remained at that low rate since. This leach rate is a factor of $10^3$-$10^6$ lower than that measured on waste glasses in dynamic laboratory leach tests, such as the proposed IAEA standard test. In fact, the initial leach rate of the Canadian glass, as measured in the laboratory, was approximately $1 \times 10^{-5}$ g glass/cm$^2$-day (27). It appears that in the actual geologic exposure system, a relatively stagnant layer of naturally occurring and leached ions exists around the glass blocks, which is inhibiting further leaching to a very low rate.

Effect of pH. Most natural waters have a pH between 4 and 9. The pH of the solution in contact with waste glass is frequently increased by the leaching reaction to approximately pH 9, the highest pH extreme observed in nature. Therefore, high pH natural waters are not a concern since they do not appreciably change the conditions that would normally arise at the surface of waste glass being leached. In extreme cases, acidic natural waters can potentially have a significant effect. At pH 4, the leach rate of most waste glasses is increased only slightly over that at pH 7, but at pH 3 the increases can be a factor of 2 to 10, or even more, depending on the glass formulation (6,7).

Borosilicate waste glasses are appreciably soluble in mineral acids (1 to 6N) because of their relative low silica content. Their solubility in mineral acids is possibly advantageous because it assures that the solidification can be reversed if, at any time in the future, it is desired to change the waste form or extract some component of the waste which has become valuable.

Effect of Salt Brine. Deep salt deposits which underlay extensive areas in many parts of the world are leading candidates for the location of final disposal sites for nuclear waste. They are obviously free of moving water, otherwise the salt would have migrated away as saline water long ago. Salt deposits do contain small amounts of stagnant waters as saturated salt brine. The total water content is usually less than 0.5% by weight. Most of the brine is in small "bubbles" occluded in the salt crystals. The brine bubbles will migrate toward heat sources because the crystalline salt is more soluble on the warm side of the bubble than on the colder side. Thus canisters of waste buried in salt formations may partially, at least, contact brine solution. It has been estimated that a maximum of 30ℓ of brine could migrate to any given canister (28). (Significant amounts of this brine may evaporate when the repository is ventilated during filling.)

More tests are being made, but the tests made thus far indicate that the leach rate of waste glass at 25°C in salt brine, or in more dilute salt water, such as seawater, is up to a factor of 10 less than the leach rate of the same glass in deionized water. Leach rates at elevated temperatures may be more pertinent since some exposure of the waste glass to salt brine is expected to occur while the repository is still at an elevated temperature. This is because rapid stress corrosion cracking that will fully penetrate the walls of the canisters is expected to occur only a few months after waste burial. Autoclave experiments are being carried out at elevated temperatures to evaluate the behavior in these circumstances (26).

Effect of Nonsaline Geologic Waters. Other potential geologies being considered for disposal of nuclear waste include granite, basalt, and shale formations. These differ from salt formations because the chemical and physical makeup of the geology is more complex and larger amounts of water can conceptually be present. Examination of the potential interactions between waste glass and these geologies is less advanced than it is with salt; only limited data has been reported (29). However, based on available data, it is expected the leach rates of waste glass in these natural waters will be lower than the leach rate of the same waste glass in deionized water. Thus the results obtained with deionized water yield conservative results when applied to the analysis of nonsaline geologic repositories. Hydrothermal reactions between waste glass and nonsaline repository materials are also being studied (29).

## Extrapolation of Leach Data--Fitting to Models

It is not practical to run leach tests for longer than a few months, or perhaps a few years--tests for process control purposes should not take more than a few days--yet leaching behavior over much longer periods should be understood. The periods of concern correspond to the lifetimes of the radioisotopes in the waste. Therefore, mathematical models, based on postulated mechanisms and observed behavior, are being developed to extrapolate expected leach behavior to times beyond which it can be experimentally verified.

Factors which should be treated in leaching models are represented graphically in Fig. 17. The location and shape of line AB is determined by the leaching mechanism, and represents leaching from a semi-infinite medium. The location and shape of line AB is very dependent on temperature and leaching media. It is important to realize that even though line AB cannot be totally verified, it is possible to verify the portion AA' within experimentally practical times, and thus establish the major behavior pattern with good confidence. Line AC represents leaching from a finite medium. Models should be able to predict releases from different sizes (and configurations) of initial waste forms. In addition, models should factor in radioactive decay, as represented by line AD. A final refinement would factor in the effect of leachant velocity. Leaching is suppressed by a low leachant flow. THE LOW FLOW SITUATION, PROBABLY MOST REPRESENTATIVE OF A GEOLOGIC REPOSITORY, WOULD BE REPRESENTED BY LINE AE.

Several models have been proposed for extrapolating the leach behavior of solidified radioactive waste forms, including glass. Most include the effects of radioactive decay and specimen size. None include the refinement

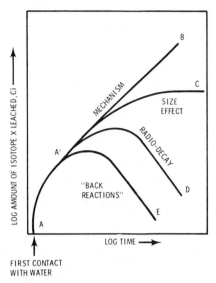

FIGURE 17. Considerations in Modeling Long-Term Leaching

of factoring in the effect of flow rate. This may be very difficult. Efforts to do it for minerals being leached by groundwater have not been successful (30).

A review of the various proposed leaching models is beyond the scope of this report. The interested reader is directed to the following list of references: Griffing (31), Cohen (32), Moore (33), Godbee (34), Logan (35), Kottwitz (36), Anders (37), Ewest (38). Of course, leaching is only part of the story; it only provides the source term. What leaching means in terms of hazard to man requires pathway analysis.

## Radiation Effects

Perhaps the most unique feature of waste glass is the high levels of contained radioactivity potentially possible. Depending on the type and amount of radioactivity incorporated in waste glass, the total beta-gamma dose to the glass may reach $10^{12}$ rads in the first one thousand years; the alpha events due to decay of the contained actinides, many of which have long half-lives, may exceed $10^{19}/cm^3$ of glass in 250,000 years.

Will self-radiation for thousands of years change the properties of waste glass in any way, making the assumptions concerning the glass used in various long-term behavior analyses invalid? It must be proven that, once the thermal and hydrothermal devitrification effects are defined, the glass will no longer change, and that long-term radiation effects can be discounted. With some reservations, described in the following sections, all evidence indicates that it is indeed legitimate to discount radiation effects. The evidence shows waste glass to be very radiation-resistant material.

Radiation can potentially cause:

• stored energy buildup;
• bulk or localized density changes;
• changes in rates of devitrification;
• amorphization of crystalline species (metamictization);
• changes in mechanical strength;
• changes in leach rate;
• helium buildup (from alpha decay).

The effects can be due to the energetics of the decay event itself or, in some cases, to the fact that the alpha and beta decays transmutate one element into another which may have a different valence, ionic radius, and chemical behavior.

Although the list of possible radiation effects is long, the tests performed to date have revealed no effects which would jeopardize the use of glass for the long-term immobilization of up to at least $1 \times 10^5$ Ci of fission products per liter of glass and $1 \times 10^3$ Ci of actinides per liter of glass.

Most radiation effects are temperature-sensitive; most of the effects diminish as temperature increases. This means that there is an inverse time relationship between the thermal effects and the radiation effects. Whereas the potential for thermal effects is only in the early life of the waste glass, the potential for radiation effects reaches a maximum in the later life of the waste glass, when it is completely cool.

## Comparison of Types of Radiation and Experimental Techniques

Waste glass will be subjected to a mixture of gamma, beta, neutron, and alpha radiation. Except for the associated heat generation, the effects of gamma and beta radiation on waste glass are minimal. This is according to theory and was demonstrated in the WSEP program. The WSEP waste glasses received doses up to $5 \times 10^{11}$R with little effect. The contained isotope responsible for the major portion of the doses received by the WSEP waste glasses was $^{144}$Ce-Pr, a combined beta-gamma emitter. Neutrons also contribute little to the potential radiation damage mechanisms for waste glass, because the neutron flux associated with $\alpha$-n reactions and spontaneous fission is comparatively low (39). The alpha particles from decay of the actinides, and particularly the recoil particles associated with alpha decay, have the major potential for creating radiation effects in waste glass (40). Therefore, it is the effects of actinides in waste glass that are receiving emphasis in studies being conducted at several sites in the U.S. and elsewhere.

The actinide content of waste glass is made up of a mixture of uranium, neptunium, plutonium, americium, and curium isotopes, plus their decay products. The relative amounts of the actinides in a waste glass depends on the source of the radioactive waste incorporated in the waste glass. In all cases, however, the actinide alpha-decay effects will continue for long times. Table 4 lists the alpha decays during the first million years for one representative waste glass.

## TABLE 4

### Time Distribution of Alpha Decay in a
### Representative Waste Glass

| Time of Vitrification | Alpha Events, Percent of Total |
|---|---|
| First 100 yr. | 3.4 |
| 100-1,000 yr. | 3.4 |
| 1,000-10,000 yr. | 6.8 |
| 10,000-100,000 yr. | 16.7 |
| 100,000-1,000,000 yr. | 69.7 |

It is apparent that the effects of the actinides cannot be studied when glasses are simply prepared with representative actinide concentrations. The doses achieved in practical experimental times would be relatively insignificant in terms of the overall dose, so techniques to accelerate the effects must be used.

Two basic accelerating techniques are available: external bombardment and internal bombardment. External bombardment with protons, alpha particles, or heavier particles can yield a very high dose rapidly, but since the effects are concentrated in a surface layer less than 100 µm deep, they are difficult to interpret. Fast neutron bombardment can give "knock on" particles with effects somewhat analogous to alpha recoil particles, but quantification of the effects is difficult because of the assumptions that must be made. Slow neutron bombardment can be used to give high energy fission fragments internally, but again somewhat tenuous assumptions have to be made to interpret the results. The conclusion is that the most realistic simulation of the actual situation is accomplished by internal bombardment using alpha decay. This is the approach that has been used in most waste glass studies.

Alpha effects are accelerated experimentally by preparing waste glasses containing actinide isotopes with short half-lives. Isotopes which have been used include $^{238}$Pu(t 1/2 = 87 years), $^{242}$CM(t 1/2 = 162 days), and $^{244}$Cm(t 1/2 = 18 years). The question then arises: What total dose is required in the accelerated experiments? The answer is simplified, because radiation effects tend to reach an equilibrium or saturation value, which changes very little with additional radiation doses. Thus for many of the effects, the goal can be simply to define the saturation values. The saturation doses can then be related to the doses expected in waste glass.

Representative doses have been calculated and reported (41). The cumulative alpha dose in glasses containing once-through LWR fuel cycle waste and full-recycle LWR fuel cycle waste will be about $3x10^{18}$ and $3x10^{19}$ α/g, respectively, after 100,000 years. Experiments have shown that most of the radiation effects reach an equilibrium, or saturation value, which is relatively independent of waste glass composition (6,7). However, the dose required to saturate different property effects does vary. For instance, density change saturates at less than $2 \times 10^{18}$ $\alpha/g$, whereas stored energy does not saturate until the cumulative dose is about $4 \times 10^{18}$ $\alpha/g$. It has been calculated that a cumulative dose of $1.26 \times 10^{19}$ $\alpha/g$ is required to achieve saturation radiation effects, i.e., full metamictization, of naturally-occurring zircon (42). (Metamictization will be discussed further in the section entitled Radiation Effects on Devitrification and Metamictization.) Cumulative dose rates exceeding $1 \times 10^{19}$ $\alpha/g$ of glass have been achieved in a waste glass at PNL which contains 8 wt % $^{244}Cm$ (6,7). Other specimens have been prepared containing 1 and 3 wt % $^{244}Cm$ so that dose rate correlations can be made. No dose rate effects have been found, giving confidence that the results obtained in the accelerated tests are valid and can be used to predict the radiation effects on waste glasses many years into the future.

The radiation effects described in the upcoming sections were all determined using internal bombardment techniques, i.e., actinide doping. They are considered valid with one reservation. In the tests to date, the actinides have been fairly uniformly distributed in the glass matrix. In devitrified waste glass, particularly, it is conceptually possible that the actinides could become concentrated to form localized areas with higher dose rates than the surrounding glass. It has already been determined that this does not occur with curium. The behavior of other actinides in devitrified glass is still being evaluated.

Stored Energy

Stored energy is latent energy stored in a substance by radiation-induced displacement of atoms from their normal lattice position. The latent energy is released as heat when the temperature of the substance is raised sufficiently. The specific heat of waste glasses is usually about 0.2 cal/°C-g. The maximum stored energies in waste glasses are in the range of 30 to 60 cal/g, thus under adiabatic conditions, the maximum additional temperature rise that could occur due to rapid release of the stored energy is only 150 to 300°C. It may be concluded that the stored energy levels in waste glass are not a hazard.

Stored energy buildup is an inverse function of temperature, as shown in Fig. 18a. Figure 18b shows that the release of stored energy is fairly uniform over a temperature range from 150°C up to the softening point of the glass, at which temperature all of the stored energy has been released. The formation of stored energy reaches an equilibrium value at 25°C after an integrated dose of less than $4 \times 10^{18}$ $\alpha/g$ of glass, as shown in Fig. 18c.

The data shown in Figs. 18a, b, and c were obtained on a specific zinc borosilicate waste glass composition using 1 and 8 wt % $^{244}Cm$. Stored energy measurements on the other waste glasses have yielded similar results within a factor of 2 to 3 (6,7,43,44). The differences are attributed to compositional effects.

Density Change

Density change is a commonly observed radiation effect in ceramic substances. The density changes observed in waste glasses, which are a maximum of about 1%, are much smaller than those observed in some very stable ceramic materials. Density changes of 14% occur in irradiated crystalline silica (45) and 8% in irradiated alumina (46).

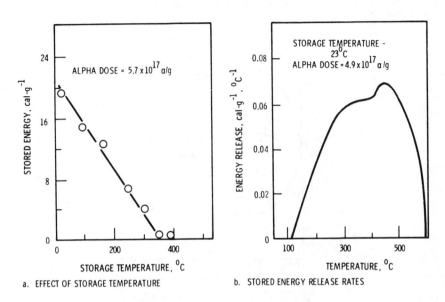

a. EFFECT OF STORAGE TEMPERATURE

b. STORED ENERGY RELEASE RATES

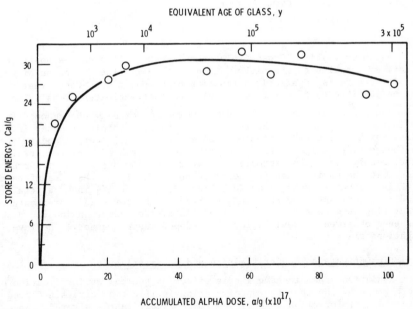

c. EFFECT OF ALPHA DOSE

FIGURE 18. Stored Energy in Waste Glass

The density change in waste glass reaches an equilibrium value after about $2 \times 10^{18}$ $\alpha$/g. Only minor density fluctuations occur with further increases in radiation dose.

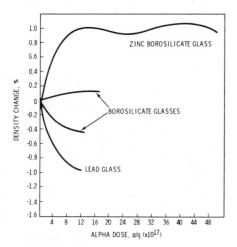

**FIGURE 19.** Radiation-Induced Density Change in Waste Glass

As shown in Fig. 19 density change in waste glass can be positive or negative; the glasses can either shrink or swell. The direction of the density change is apparently a function of composition. Conceptually, it may be possible to formulate waste glasses that exhibit no density change.

Two potential effects of radiation-induced density change need to be evaluated: 1) possible canister strain, and 2) possible stresses in the glass which will cause some spontaneous fracturing. Tests indicate that neither effect is likely to cause problems. Increased canister strain results only with glass formulations where density decreases with radiation. At the maximum density decrease observed of 1%, the circumference of a 12-in.-diameter canister would be increased about 1/4 inch, a strain readily accommodated by the canister metal. The potential for spontaneous fracturing is greatest in devitrified glass. No radiation-induced cracking has been observed in devitrified $^{244}$Cm-doped waste glasses. The investigations are continuing.

## Radiation Effects on Devitrification and Metamictization

Radiation seems to have little effect on thermal devitrification of waste glasses. Neither the species of devitrification crystals formed nor their rate of formation is believed to be significantly altered. Specimens of four different PNL glass compositions have been prepared with full levels of radioactivity. Similar specimens, but with nonradioactive isotopes of the fission products and chemical standins (rare earths) for the actinides, have also been prepared. The nonradioactive and radioactive specimens have been subjected to the same thermal devitrification conditions. Preliminary results confirm expectations; devitrification behavior is similar. Radiation would not be expected to affect thermal devitrification since radiation effects are annealed out at the temperatures where the devitrification rate is significant (above 500°C).

Over the long term, however, radiation can potentially be expected to affect some crystals in thermally-devitrified glass. This conclusion is based on evidence from the metamict minerals. Metamict minerals are naturally-occurring minerals which, when examined by x-ray diffraction, are found to be amorphous (47). When heated they release stored energy and become crystalline. The metamict minerals were probably originally crystalline and became amorphous over many millions of years due to radiation effects from small amounts of contained uranium and/or thorium. This is why over the long term, the contained radioactivity in waste glass may tend to reverse devitrification by converting some crystalline phases back to an amorphous glass-like state.

## Radiation Effects on Leach Rates

Leach tests made on waste glasses containing up to $9 \times 10^4$ Ci/$\ell$ of fission products are not more than a factor of two higher than those made on nonradioactive simulated waste glasses of the same general chemical composition (48,49). This difference is within the normal experimental deviation of leach test data. Similarly, the leach rate of a $^{244}$Cm-spiked waste glass was not significantly different from the leach rate of a comparable nonradioactive glass after a cumulative dose over $3 \times 10^{17}$ $\alpha$ disintegrations/g (41). Thus radiation does not affect the leach rate of waste glasses, and leach data obtained on nonradioactive simulated waste glass can be used to predict the leaching behavior of fully radioactive waste glass. Tests are producing more data which confirm this conclusion.

In some instances, the internal surfaces of glass vessels containing highly alpha-active solutions have become crazed and have even begun to slough off flakes of surface material (50). In those cases, the actinide concentrations in solution have been high ($>3 \times 10^{-4}$ Ci/m$\ell$), probably higher than would exist in any solution likely to contact waste glass, even after prolonged static contact. This is an area that needs further investigation, although the effect is not thought to have an important impact even if it occurs. For instance, Scheffler et al. (43,44) studied waste glasses having an abnormally high plutonium content (2 wt %) and observed a periodic sloughing off of a hydrous silica-rich surface layer. Even so, the leach rate was low, less than $1 \times 10^{-7}$g of glass/cm$^2$-day at 25°C. The small amount of plutonium leached could be filtered; it was evidently tightly bonded to hydrous silica particles.

## Helium Behavior

Helium atoms are generated wherever alpha decay events occur in waste glass. Each alpha particle captures two electrons to become a helium atom. Helium is slightly soluble in glass; helium also diffuses through glass

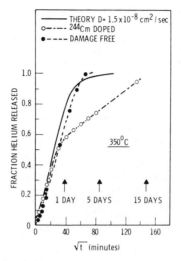

**FIGURE 20.** Isothermal Helium Release From Glass

quite readily. The diffusion is retarded as the glass composition increases in complexity; thus, helium diffusion in waste glass is lower than in quartz glass or commercial soda-lime-silica glass (51). Figure 20 shows that the diffusion of a portion of the helium produced by alpha decay is also further inhibited, when compared to that of helium implanted in the glass by solubility at high temperatures. The inhibition is attributed to trapping of some of the alpha-produced helium at point defects. Expressions for both untrapped ($D_u$) and trapped ($D_t$) helium diffusion coefficients have been experimentally derived (50)

$$D_u = 2.1 \times 10^{-3} \exp(-15,000/RT) \ cm^2/sec,$$

$$D_t = 1.7 \times 10^{-4} \exp(-15,000/RT) \ cm^2/sec.$$

Using these diffusion coefficients, canister pressurization was estimated at two temperatures, 25 and 300°C, as a function of time for waste glass containing LWR once-through and full recycle wastes. The estimated pressures are plotted in Fig. 21a and b. The solid lines represent no trapped helium, the shaded portion below the lines represents the potential effect of trapped helium. A 10 vol % plenum in the canister and 30 wt % waste loading in the glass was assumed.

Defense waste glasses are not shown in the figures since their helium generation rate is orders of magnitude below that of the LWR fuel cycle waste glasses. The important point is that helium pressurization is negligible even in LWR waste glass canisters; only in the unlikely event that the full recycle waste glass canisters will still be at 300°C after 100 years, could the pressure reach 150 psig.

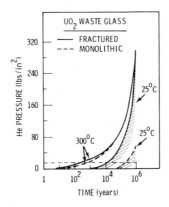

a.  Once-Through

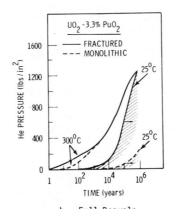

b.  Full Recycle

**FIGURE 21.** Potential Helium Pressure in Waste Glass Canisters

## Transmutation

Radioactive decay results in the transmutation of the decaying isotope, often to atoms of significantly different valence and ionic radius.  Some of the transmutations that occur in waste glass are shown in Table 5.  Intralanthanide and actinide transmutations are now shown because the resultant changes are insignificant.  The actual number of transmutations that will occur in a given waste glass depends on the fission product loading and the age of the waste when the glass is made.  Even in glasses with a high waste loading, the number of nonoxygen atoms undergoing transmutations will not exceed 2%. Because of the nonstoichiometric nature of the glass, it is expected that the transmutations can be accommodated without effect. The exception may be severely devitrified glass, because the effect of transmutation on the stability of crystals could be more severe.

## References

1. S.C. Slate and R.F. Maness, Corrosion Experience in Nuclear Waste Processing at Battelle-Northwest, Mat. Performance 17, 13-21 (June 1978).
2. D.J. Bradley (Nov. 1977), History of Prototype High-Level Waste Canister SS-9 While in Air and Water Storage, PNL-2278, Battelle, Pacific Northwest Laboratories, Richland, WA.
3. R.M. Elliot, Glass Composition and Density Changes, J. Amer. Ceram. Soc. 28, 303-305 (1945).
4. T.H. Smith and W.A. Ross (May 1975), Impact Testing of Vitreous Simulated High-Level Waste in Canisters, BNWL-1903, Battelle, Pacific Northwest Laboratories, Richland, WA.
5. L.B. Bunnell (Nov. 1977), Mechanical and Thermal Shock, Quarterly Progress Report, Research and Development Activities, Waste Fixation Program, January Through March 1977, PNL-2265-1, pp. 23-29, Battelle, Pacific Northwest Laboratories, Richland, WA.
6. W.A. Ross (June 13, 1978), Process for Solidifying High-Level Nuclear Waste, U.S. Patent No. 4 094,809.
7. W.A. Ross et al. (June 1978), Annual Report on the Characterization of High-Level Waste Glasses, PNL-2625, Battelle, Pacific Northwest Laboratories, Richland, WA.
8. A.K. De, B. Luckscheiter, G. Malow, and E. Schweiwer (Sept. 1977), Fission Products in Glasses, Part II: Development of Glass Ceramics, HMI-B218, Hahn-Meitner Institute, Berlin, Germany.
9. R.J. Thompson, J.E. Mendel, and J.H. Kleinpeter (June 1970), Waste Solidification Demonstration Program: Characterization of Nonradioactive Samples of Solidified High-Level Waste, BNWL-1393, Battelle, Pacific Northwest Laboratories, Richland, WA.
10. R.P. Turcotte and J.W. Wald (March 1978), Devitrification Behavior in a Zinc Borosilicate Waste Glass, PNL-2247, Battelle, Pacific Northwest Laboratories, Richland, WA.
11. G. Malow and E. Schiewer (Sept. 1972), Fission Products in Glasses, Part I: Borosilicate Glasses Containing Fission Products, HMI-B217, Hahn-Meitner Institute, Berlin, Germany.
12. R.H. Doremus (1973), Phase Separation, Glass Science, John Wiley and Sons, NY, pp. 44-73.
13. DPST-76 (1976), Savannah River Laboratory Quarterly Report, Waste Management, July-September 1976, DPST-76-125-3, Savannah River Laboratory, Aiken, SC.

14. B.S. Shukla, G.A. Vaswani, and N.S. Sunder Rajan (1977), Studies on the Immobilization of Sulphate-Bearing Waste, PBO-$B_2O_3$-$SiO_2$ System, BARC-951, Bhabha Atomic Research Centre, Bombay, India.
15. R. Bonniaud (Nov. 1966), Survey of the Studies Conducted in France on the Solidification of Concentrated Fission Product Solutions, Proc. Symp. Solidification and Long-Term Storage of Highly Radioactive Wastes, CONF-660208, U.S. Atomic Energy Commission, pp. 120-138.
16. L. Zager and A. Schillmoeller, The Physical and Chemical Processes Involved in the Leaching of Glass Surfaces by Water, Glastechnische Berichte 33, 109-116 (1960); available in English as AERE Trans. 867, Atomic Energy Research Establishment, Harwell England (March 1961).
17. A. Paul, Chemical Durability of Glasses: A Thermodynamic Approach, J. Mat. Sci. 12, 2246-2268 (1977).
18. P.B. Adams and J.K. Lunney, Water for ASTM Test C-225: Chemical Durability of Glass Bottles, Ceram. Bull. 49, 543-548 (1970).
19. F.R. Bacon, The Chemical Durability of Silicate Glass, Glass Industry (Aug., Sept., Oct. 1968).
20. E.D. Hespe, Leach Testing of Immobilized Radioactive Waste Solids, Atomic Energy Review 9, 195-207 (1971).
21. J.E. Mendel and I.M. Warner (April 1973), Soxhlet Leach Test, Quarterly Progress Report, Research and Development Activities, Waste Fixation Program, December 1972 Through March 1973, BNWL 1741, pp. 6-7, Battelle, Pacific Northwest Laboratories, Richland, WA.
22. P.B. Adams (1972), Glass Containers for Ultra-pure Solutions, Ultrapurity, Marcel Dekker, NY, pp. 293-351.
23. S.M. Budd, The Mechanism of Chemical Reaction Between Silicate Glass and Attacking Agents, Part 1, Electrophilic and Nucleophilic Mechanisms of Attack, Phys. Chem. Glasses 2, 111-114 (1961).
24. S.M. Budd and J. Frankiewicz, Part 2, Chemical Equilibria at Glass-Solution Interfaces, Phys. Chem. Glasses 2, 115-118 (1961).
25. J.E. Mendel, J.L. McElroy, and A.M. Platt (1976), High-Level Radioactive Waste Management Research and Development Program at Battelle-Pacific Northwest Laboratory, High-Level Radioactive Waste Management, Ed. M.H. Campbell, American Chemical Society, Washington, DC, pp. 93-107.
26. J.H. Westsik, Jr., and R.P. Turcotte (Sept. 1978), Hydrothermal Reactions of a High-Level Waste Glass in Salt Brine and Deionized Water, PNL-2759, Battelle, Pacific Northwest Laboratories, Richland, WA.
27. W.F. Merritt, High-Level Waste Glass: Field Leach Test, Nucl. Tech. 32, 88-91 (1977).
28. G.H. Jenks (1972), Radiolysis and Hydrolysis in Salt Mine Brines, ORNL-TM-3717, Oak Ridge National Laboratory, Oak Ridge, TN (1972).
29. G.J. McCarthy, W.B. White, R. Roy, B.E. Scheetz, S. Komarneni, D.K. Smith, and D.M. Roy, Interactions Between Nuclear Waste and Surrounding Rock, Nature 273, 217-219 (1978).
30. T. Paces, Steady-State Kinetics and Equilibrium Between Groundwater and Granitic Rock, Geochim. Cosmochim. Acta 37, 2641-2663 (1973).
31. D.E. Griffing (Aug. 1974), Progress Report on the Development of Models for the Description of Radionuclide Transport in Solids, ARH-SA-190 REV, Atlantic Richfield Hanford Company, Richland, WA 99352 (Aug. 1974).

32. J.J. Cohen et al. (July 1977), Determination of Performance Criteria for High-Level Solidified Nuclear Waste, NUREG-0279.

33. J.G. Moore, H.W. Godbee, and A.H. Kibbey, Leach Behavior of Hydro-fracture Grout Incorporating Radioactive Wastes, Nucl. Tech. 32, 39-52 (1977).

34. H.W. Godbee and D.S. Joy (1974), Assessment of the Loss of Radioactive Isotopes from Waste Solids to the Environment, Part I: Background and Theory, ORNL-TM-4333, Oak Ridge National Laboratory, Oak Ridge, TN.

35. S.E. Logan and M.C. Berbane (1977), Geologic Modeling in Risk Assessment Methodology for Radioactive Waste Management, Workshop Proceedings, ISPRA, Risk Analysis and Geologic Modeling in Relation to the Disposal of Radioactive Wastes into Geologic Formation, OECD Nuclear Energy Agency, May 23-27, pp. 77-115.

36. D.A. Kottwitz (July 1974), Mathematical Model for Leaching of Glass by Water, Quarterly Progress Report, Research and Development Activities, Waste Fixation Program, April Through June, 1974, BNWL-1841, pp. 53-55, Battelle, Pacific Northwest Laboratories, Richland, WA.

37. O.V. Anders, J.F. Bartel, and S.J. Altschuler, Determination of the Leachability of Solids, Anal. Chem. 50, 564-569 (April 1978).

38. E. Ewest and H.W. Levi (1976), Evaluation of Products for the Solidification of High-Level Radioactive Waste from Commercial Reprocessing in the Federal Republic of Germany, Proc. of Symp. on Management of Radioactive Waste from the Nuclear Fuel Cycle, II, International Atomic Energy Agency, Vienna, Austria (March 22-26).

39. D.F. Newman (April 1973), Radiation Damage in Borosilicate Glass, Quarterly Progress Report, Research and Development Activities, Waste Fixation Program, December 1972 Through March 1973, BNWL-1741, pp. 30-38 Battelle, Pacific Northwest Laboratories, Richland, WA.

40. G.H. Jenks and C.D. Bopp (Jan. 1973), Energy Storage in High-Level Radioactive Waste and Simulation and Measurement of Stored Energy with Synthetic Wastes, ORNL-TM-3781, Oak Ridge National Laboratory, Oak Ridge, TN.

41. J.E. Mendel, W.A. Ross, F.P. Roberts, Y.B. Katayama, J.H. Westsik, Jr. R.P. Turcotte, J.W. Wald, and D.J. Bradley (June 1977), Annual Report on the Characteristics of High-Level Waste Glasses, BNWL-2252, Battelle, Pacific Northwest Laboratories, Richland, WA.

42. L.A. Bursill and A.C. McLaren, Transmission Electron Microscope Study of Natural Radiation Damage in Zircon ($ZrSiO_2$), Phys. Stat. Sol. 13, 331-343 (1966).

43. K. Scheffler and V. Riege, Investigations on the Long-Term Radiation Stability of Borosilicate Glasses Against Alpha-Emitters, KFK 2422, Kernforschungszentrum, Karlsruhe, Germany (April 1977).

44. K. Scheffler, V. Riege, K. Louwrier, Hj. Matzke, I. Ray, and H. Thiele, Long-Term Leaching of Silicate Systems: Testing Procedure, Actinide Behavior and Mechanisms, KFK-2456, Kernforschungszentrum, Karlsruhe, Germany (1977).

45. E. Lell, N.J. Kriedl, and J.R. Hensler, Radiation Effects in Quartz, Silica, and Glasses, Prog. Ceram. Sci. 4, 1-93 (1966).

46. F.W. Clinard, Jr., J.M. Bunch, and W.A. Ranken (1975), Proc. Intl. Conf. Radiation Effects and Tritium Technology for Fusion Reactors, II, CONF-750989, Gatlinburg, TN, p. 498.

47. A. Pabst, Am. Mineral. 37, 137 (1952).

48.  J.E. Mendel and J.L. McElroy (July 1972), Waste Solidification
     Program, Volume 10, Evaluation of Solidified Waste Products, BNWL-
     1666, Battelle, Pacific Northwest Laboratories, Richland, WA.
49.  J.E. Mendel (July 1973), Measurements on Core-Drilled Samples,
     Quarterly Progress Report, Research and Development Activities, Waste
     Fixation Program, April Through June 1973, BNWL-1761, pp. 16-18,
     Battelle, Pacific Northwest Laboratories, Richland, WA.
50.  D.G. Tuck, Radiation Damage to Glass Surfaces by $\alpha$-Particle Bombard-
     ment, Intl. J. Appl. Rad. and Isotopes 15, 49-57 (1964).
51.  R.P. Turcotte (May 1976), Radiation Effects in Solidified High-Level
     Wastes, Part 2 -Helium Behavior, BNWL-2051, Battelle, Pacific
     Northwest Laboratories, Richland, WA.

APPENDIX II

MINERAL MODELS FOR CRYSTALLINE HOSTS FOR RADIONUCLIDES

by

Gregory J. McCarthy, The Pennsylvania State University
William B. White, The Pennsylvania State University
Deane K. Smith, The Pennsylvania State University
Antonio C. Lasaga, The Pennsylvania State University
Rodney C. Ewing, University of New Mexico
Alastair W. Nicol, University of Birmingham, England
Rustum Roy, The Pennsylvania State University

Reproduced from Draft Environmental Impact Statement
"Management of Commercially Generated Radioactive Wastes"
DOE/EIS/0046-D

Appendix II.  STABILITY OF MINERALS THAT COULD CONTAIN RADIONUCLIDES

This section presents a review of minerals that are candidate hosts for the radionuclides discussed in Chapter 7.  They are either known to contain substantial amounts of these elements or they are likely to accept these elements based on crystal chemical predictions.  The physico-chemical and crystal chemical criteria for selecting host minerals, along with the common mineral synthesis methods, are discussed and tables of candidates are presented.  A thorough treatment of what is known about the process of metamictization and metamict minerals is also included.

## PHYSICO-CHEMICAL PRINCIPLES

### Stability Criteria

#### Use of Solubility Data

One of the first tasks is to lay out clearly the physical and chemical foundations used to define whether a known mineral is classified as very stable, relatively unstable, or very unstable with respect to alteration, weathering and diagenesis.  Chemical weathering and alteration are most often the result of the interaction between an electrolyte aqueous solution and the various minerals being weathered.

There are several factors which play an important role in determining the mobility of elements via weathering ionic solutions.  One group of factors is related to the overall physical properties of the "weathering system," i.e. of the hydrologic system and the host mineral assemblage.  For example, the flow rate of solution through a permeable system is determined by Darcy's Law:

$$\vec{u} = -\frac{k}{\mu} (\rho \vec{g} + \nabla P) \qquad \vec{v} = \frac{\vec{u}}{\rho \emptyset}$$

where
$\vec{u}$ = fluid flux vector $(g/cm^2/sec)$
$\vec{v}$ = true fluid velocity (cm/sec)
$\vec{g}$ = gravity force vector $(cm/sec^2)$
$k$ = permeability of the rock assemblage $(cm^2)$
$\mu$ = viscosity of the fluid $(cm^2/sec)$
$\rho$ = density of the fluid $(g/cm^3)$
$P$ = pressure (bars)
$\emptyset$ = porosity of rock.

Clearly, then, the water flow depends on gravity and the pressure gradient at the given locality (a property of the hydrologic system as a whole) as well as on the porosity and permeability of the rock assemblage in the locality and the density and viscosity of the fluid.

The hydrodynamic equations, which incorporate Darcy's Law, allow us to calculate the hydrodynamic mobility of a given cation or anion in solution from its original location within a given mineral of the weathered rock to some other depositional place such as a sedimentary deposit, rivers, oceans or the biosphere.  However, we can obtain absolute flux rates for a given ion (i.e. moles/$cm^2$/sec) only if we know its concentration in the percolating solution.

The magnitude of the concentration of a given element in a solution, which is in contact with a weathering mineral assemblage, is the central element used in establishing the intrinsic stability of a particular nuclear waste element-containing mineral to alteration and weathering. This concentration is generally a function of time, since it is kinetically controlled. Nevertheless, almost all geochemical work on the mobility of elements via solutions has applied a thermodynamic and not a true kinetic approach. Whether true thermodynamic equilibrium is reached between solution and a particular mineral depends, among other things, on how long they are in contact (i.e. the flow rate), a concept which often appears as the ambiguous "water-rock ratio" in the literature. It seems likely that under most circumstances the concentration of an element in a weathering solution will be kinetically controlled. Unfortunately, there is a dire need for suitable kinetic data. The kinetic factors involved in the time dependence of the concentration, which may keep the concentration well below the thermodynamic limit, will be discussed below.

One can usually establish only an upper limit to the concentration of a given element by the use of thermodynamics. Assuming equilibrium between minerals and solution, the concentration of any particular nuclear waste element will then be governed by the solubility of the minerals containing it.

Before discussing the thermodynamic approach to stability, a brief review of the general qualitative work on weathering stability in the literature is presented. Soil geochemists have set up a qualitative scale of the different inherent tendencies of minerals to alter by weathering processes. The weathering rate depends on the structure and composition of the minerals, as well as the weathering environment. Goldich[1] formulated such a weathering stability series for the major elements. He found that the major elements are removed from rocks and minerals in the order:

$$Ca^{+2} > Na^+ > Mg^{++} > K^+ > SiO_2 > Fe_2O_3 > Al_2O_3.$$

Loughnan[2] gives a similar result (see Table P.1).

### TABLE P.1. Mobilities of the Common Cations

↑ Increasing Rate of
Loss from the Environment →

1. $Ca^{++}$, $Mg^{++}$, $Na^+$ - readily lost under leaching conditions.
2. $K^+$ - readily lost under leaching conditions but rate may be retarded through fixation in the illite structure.
3. $Fe^{++}$ - rate of loss dependent on the redox potential and degree of leaching.
4. $Si^{4+}$ - slowly lost under leaching conditions.
5. $Ti^{4+}$ - may show limited mobility if released from the parent mineral as $Ti(OH)_4$; if $TiO_2$ form, immobile.
6. $Fe^{3+}$ - immobile under oxidizing conditions.
7. $Al^{3+}$ - immobile in the pH range of 4.9-9.5.

Much less is known about the relative mobilities of the trace elements (Ln, actinides, etc.). Jackson and his colleagues[3-5] have set up a weathering sequence of clay-size minerals in soils and sedimentary deposits (see Table P.2). Pettijohn[6] has compared the frequency of occurrence of each species in recent and older sediments and established an order of persistence, which is in agreement with the Goldich series (see Table P.3).

TABLE P.2. Weathering Sequence of Clay-Size Minerals in Soils and Sedimentary Deposits[a]

| Weathering stage and symbol | Clay-size mineral occurring at various stages of the weathering sequence |
|---|---|
| 1, Gp | Gypsum (also halite, etc.) |
| 2, Ct | Calcite (also dolomite, aragonite, etc.) |
| 3, Hr | Olivine-hornblende (also diopside, etc.) |
| 4, Bt | Biotite (also glauconite, chlorite, antigorite, etc.) |
| 5, Ab | Albite (also anorthite, microcline, stilbite, etc.) |
| 6, Qtz | Quartz (also cristobalite, etc.) |
| 7, Il | Illite (also muscovite, sericite, etc.) |
| 8, X | Hydrous Mica - Intermediates |
| 9, Mt | Montmorillonite (also beidellite, etc.) |
| 10, Kl | Kaolinite (also halloysite, etc.) |
| 11, Gb | Gibbsite (also boehmite, etc.) |
| 12, Hm | Hematite (also goethite, limonite, etc.) |
| 13, An | Anatase (also rutile, ilmenite, corundum, etc.) |

a. After Jackson, et al.[3]

TABLE P.3.  Persistence Order of Minerals[a,b]

| | | | |
|---|---|---|---|
| -3. | Anatase | 10. | Kyanite |
| -2. | *Muscovite* | 11. | Epidote |
| -1. | Rutile | 12. | *Hornblende* |
| 1. | Zircon | 13. | Andalusite |
| 2. | Tourmaline | 14. | Topaz |
| 3. | Monazite | 15. | Sphene |
| 4. | Garnet | 16. | Zoisite |
| 5. | *Biotite* | 17. | *Augite* |
| 6. | Apatite | 18. | Sillimanite |
| 7. | Ilmenite | 19. | Hypersthene |
| 8. | Magnetite | 20. | Diopside |
| 9. | Staurolite | 21. | Actinolite |
| | | 22. | *Olivine* |

a. After Pettijohn[6].

b. Italics signify common minerals listed in the Goldich sequence.

Although still poorly understood, structure must play an important part in the accessibility of waters to the soluble cations. Thus orthosilicates, e.g. olivine, weather much faster than framework silicates, e.g. feldspars and quartz. However, zircon, also an orthosilicate, is highly resistant to weathering, indicating that resistance to weathering cannot be based solely on such a simple structural division of the silicates.

The qualitative lists of minerals in Tables P.2 and P.3 should be quantitatively understood in terms of both thermodynamics (i.e. solubility data) and kinetics (i.e. leaching rates). The solubility and hence the thermodynamic stability of a particular mineral in a weathering solution depends on many environmental factors such as pH, Eh, complexing agents, temperature, fixation/adsorption, ion exchange and ionic strength, as explained briefly below:

- **pH:** Most minerals are leached faster and have higher solubilities in acid environments. The natural range of possible weathering solutions is pH = 4-10. One of the earliest steps in the chemical weathering of a mineral is the exchange of the small and mobile $H^+$ ion for a cation on the mineral surface, with subsequent disruption of the structure.[2] Obviously, low pH solutions can accomplish this more effectively.

- **Eh:** For ions which can exist in several valence states (e.g. $U^{+4}$ and $U^{+6}$) Eh plays a very important role in determining their solubility. The Eh of natural solutions in contact with the atmosphere is ∿600 mv. Subsurface solutions can have an Eh range of -400 mv to +400 mv, with the more reducing (low Eh) conditions found in alkaline environments.[7] For example, a mineral with very low solubility, such as uraninite ($UO_2$), requires a low Eh for stability to weathering (i.e. Eh < +200 mv if pH = 6, Eh < 0 mv if pH = 8).[21]

- Complexing: The formation of complexes has long been recognized as essential in explaining the transport of metals required to form ore deposits. The same must be investigated for the cations of the nuclear waste elements, since complexing can increase the solubility of an element by several orders of magnitude. At lower temperatures (<200°C), we expect carbonates, phosphate, sulfate/sulfide and organic complexes to be important.

- Temperature: The solubility of various minerals can change significantly with temperature. Temperatures in the vicinity of synthetic minerals containing heat producers ($^{90}Sr$, $^{137}Cs$, An's) could rise up to several hundred degrees above ambient.

- Adsorption: The ability of ions, such as $K^+$, to adsorb strongly to clays and other minerals, retards their mobility and limits their concentration in solution, following leaching of the ions. This may be important, for example, in the case of uranium, which adsorbs strongly to Mn-oxides and Fe-oxides and hydroxides.

- Cation Exchange: An important consideration in establishing the stability of a given nuclear waste element-containing mineral to the leaching of such elements, is the ability of that mineral to exchange the troublesome nuclear waste element for another ion in solution. Thus, $K^+$ may be exchanged for $Cs^+$, or $Cl^-$ may be exchanged for $I^-$. On the other hand, ion exchange of the radionuclide with clays and other minerals can also retard the mobility of the radionuclide in solution.

Rai and Lindsay[8] applied simple solubility calculations to deduce the relation between log $a_{Al}$ and log $a_{H_4SiO_4}$ at given values of pH, T and solution compositions (e.g. $a_{Ca}$, $a_{Mg}$, etc.) for several aluminosilicates. At a given value of $a_{H_4SiO_4}$, the minerals with the lowest $a_{Al^{3+}}$ would be more stable. Using values of $a_{H_4SiO_4}$ typical of soil waters ($a_{H_4SiO_4} \sim 10^{-3.2}$ m) they obtain the stability sequence muscovite > microcline > low albite > anorthite > analcime > pyroxene > K-glass (K-feldspar composition) > Na-glass (albite composition), which is in agreement with Goldich's sequence. Likewise one can plot regions of stability for various minerals on an Eh-pH diagram, as outlined by Garrels and Christ.[7]

## Use of Geologic Data

Because geologic time spans the lifetimes of the radionuclides of the critical elements, it is very logical to use nature as a laboratory and examine conditions of stability of minerals which may contain the critical elements. In general one recognizes three main geologic environments: igneous, sedimentary, and metamorphic, and asks which mineral phases may exist in each environment and what happens to a mineral grain as it sees a change in its environment. Minerals of the igneous environment see extreme temperatures (and pressures) such that they have crystallized from a melt or a fluid derived from a melt (pegmatites and hydrothermal deposits). The sedimentary environment includes the effect of exposure to the atmosphere and running water and the physical effects of separation and movement of mineral grains. The metamorphic environment involves changing pressure, temperature and pore fluid conditions inducing mineral changes in situ.

As one identifies mineral species which may be potential repository compounds, a test of their stability is to determine the geologic environments under which they can endure. If any modifications in the mineral phase do occur, then the time frame of the modifications can also be deduced. The best test of a mineral's stability is to determine the range of changes through which it can exist.

Many of the minerals which are potentially interesting host phases form initially in the igneous environment. Feldspars, feldspathoids and micas crystallize directly from the melt. Many others are pegmatitic in origin, especially those containing rare earth elements. This information implies conditions which may be necessary to form the phase desired. It may not be the only condition under which the compound will form.

After the compound has formed, the question of what happens to it as the conditions change may be answered. Because stability is the main question, one asks what phase may endure weathering and erosion unchanged, and what new phases are formed if changes do occur. There are many minerals which survive the rigors of weathering and erosion and these are ultimately collected in detrital deposits. When the detrital deposit has an economic value it is called a placer. These minerals are usually of high density and chemical resistance. Other minerals, called detrital-heavy minerals, may not survive the entire erosion cycle but persist for quite some time. Detrital heavy minerals may last sufficiently long to allow included radionuclides sufficient time to decay. It will be useful to identify the placer minerals and other detrital-heavy minerals.

The Placer Minerals. Table P.4 identifies the minerals which have been recognized in placer deposits. These minerals are characterized by high densities and chemical and physical resistance. All the noble metals--platinum, iridium, palladium, gold--are known to occur as placer minerals. Many oxides containing Ln's as well as carbonates, phosphates, tungstates and silicates are known placer minerals and therefore are potential Ln and An phases. Some low density minerals occur in placers such as quartz, spinel, garnet, corundum and diamond.

Other minerals might well be on this list of placer minerals under special conditions. If the sedimentary conditions were more reducing than usually occurs in nature, uraninite and many sulfide minerals could survive. This possibility is evidenced by the placers

TABLE P.4. Placer Minerals[a]

Element minerals

Platinum, Osmium, Palladium, Iridium, Platiniridium, Iridosmine, Osmiridium, Ferroplatinum, Gold, Electrum, Silver, Diamond

Oxide minerals

Tantalite, $FeTa_2O_6$; Thoreaulite, $ThTi_2O_6$; Cassiterite, $SnO_2$; Samarskite, $YNb_2O_6$; Baddeleyite, $ZrO_2$; Euxenite, $YNb_2O_6$; Chromite, $FeCr_2O_4$; Magnetite, $Fe_3O_4$; Columbite, $FeNb_2O_6$; Polycrase, $YTi_2O_6$; Aeschynite, $YTi_2O_6$; Loparite, $CeTi_2O_6$; Ilmenorutile $(Ti,Nb)_3O_6$; Ilmenite, $FeTiO_3$; Zirkelite, $CaZrTi_2O_7$; Pyrochlore, $Ca_2Nb_2O_6OH$; Rutile, $TiO_2$; Brookite, $TiO_2$; Anatase, $TiO_2$; Corundum, $Al_2O_3$; Spinel, $MgAl_2O_4$; Quartz, $SiO_2$

Tungstate minerals

Ferberite, $FeWO_4$; Wolframite, $(Fe,Mn)WO_4$; Hübnerite, $MnWO_4$; Scheelite, $CaWO_4$

Phosphates

Monazite, $CePO_4$; Xenotime, $YPO_4$

Carbonates

Bastnaesite, $CeCO_3F$; Parisite, $Ce_2Ca(CO_3)_3F_2$

Silicates

Thorite, $ThSiO_4$; Zircon, $ZrSiO_4$; Garnet, $(Fe,Mg)_3Al_2Si_3O_{12}$; Topaz, $Al_2SiO_4F_2$; Phenakite, $Be_2SiO_4$

---

a. Simplified formulae are given. Actual minerals usually contain many additional solid solution substitutions.

of the Witwatersrand District of Africa which formed in the reducing environments of the Pre-Cambrian.

Detrital Minerals. A great many minerals survive long distances of transport in stream beds, although the final fraction of that mineral is often much lower than in the source area. These minerals are listed in Table P.5. The rate of degradation of some of these minerals may be sufficiently slow to allow that phase to be a host for radionuclides. Minerals such as apatite, barite, allanite and titanite are particularly interesting. Apatite and allanite contain significant amounts of Ln's and An's. Strontium varieties of apatite also occur.

Mineral Associations. In addition to defining regions of stability for specific mineral phases, geologic evidence indicates which phases may occur together in an equilibrium assemblage. These mineral associations are good indicators of compatible phases. The pegmatite environment contains many of the minerals of interest. Rare earth phosphates, rare earth oxides and rare earth carbonates which are good hosts for the Ln's and An's, coexist with a variety of complex silicates which may host other critical elements. These in turn coexist with some of the common silicates which may be more appropriate hosts for $^{90}Sr$ and $^{137}Cs$.

Kinetic Factors

Often the concentration of an element in solution is not determined by thermodynamic solubility data but by the kinetics of water-rock interactions. Data on this part of the stability criteria are most urgently needed. We will outline here the major elements in the

## TABLE P.5. Detrital minerals[a]

**Elements**

Lead

**Oxide minerals**

Hematite, $Fe_2O_3$; Uraninite, $UO_2$; Uranothorite, $(U,Th)O_2$; Leucoxene, Ti oxide-hydroxide

**Sulfide minerals**

Cinnabar, $HgS$; Pyrite, $FeS_2$; Marcasite, $FeS_2$; Chalcopyrite, $CuFeS_2$; Arsenopyrite, $FeAsS_2$; Pyrrhotite, $Fe_{1-x}S$; Molybdenite, $MoS_2$; Cobaltite, $CoAsS_2$; Dyscrasite, $Ag_3Sb$; Michenerite, $PdBiTe$; Geversite, $PtSb_2$; Glaucodot, $CoAsS$; Moncheite, $PtTe$

**Sulfate minerals**

Barite, $BaSO_4$

**Phosphate minerals**

Apatite, $Ca_5(PO_4)_3F$

**Silicate minerals**

Actinolite, $Ca_2(Fe,Mg)_5Si_8O_{22}(OH)_2$; Andalusite, $Al_2SiO_5$; Biotite, $K(Fe,Mg)_3AlSi_3O_{10}(OH)_2$; Chlorite, $(Mg,Fe)_6(Al,Si)_4O_{10}(OH)_8$; Chloritoid, $(Fe,Mg)Al_4Si_2O_{10}(OH)_4$; Hornblende, $Ca_2(Fe,Mg,Al)_5Al_2Si_6O_{22}(OH)_2$; Hypersthene, $(Mg,Fe)SiO_3$: Kyanite, $Al_2SiO_5$; Olivine, $(Mg,Fe)_2SiO_4$; Allanite, $Ce_2Al_2FeSi_3O_{11}O(OH)$; Sillimanite, $Al_2SiO_5$; Staurolite, $Fe_2Al_9Si_4O_{23}OH$; Titanite, $CaTiSiO_5$; Tourmaline, $Na(Mg,Fe)_3Al_6(BO_3)_3Si_6O_{18}(OH)_4$; Zoisite, $Ca_3Al_2Si_3O_{11}O(OH)$; Gadolinite, $Be_2Y_2FeSi_2O_{10}$

---

a. Simplified formulae are given. Actual minerals usually contain many additional solid
   solution substitutions.

treatment of the kinetic stability of various minerals.

Leaching rate. If the leaching is surface-controlled, the rate at which a cation is leached from a mineral depends on (a) the reactive specific surface area of the mineral in the solution; (b) the concentrations of the species or ions involved in the transition state of the rate-determining step for surface reaction; (c) the free energy of activation of the activated complex; and (d) the temperature of the solution-rock system. The effects of pH, Eh and complexes enter via their effect on the numbers in (b). The role of temperature in kinetic processes is much more prominent than its role in solubility calculations, due to the high activation energies (10-100 Kcal/mole) often encountered. Thus it is crucial to measure accurately the activation energies for the important leaching rates.

Leaching rates can also be controlled by the rate of transport (i.e. diffusion) of leached cations from the weathering mineral-solution interface to the bulk of the solution. In this case, temperature will play a much more minor role, since diffusion activation energies are $\sim$4-5 Kcal/mole in electrolyte solutions. Experiments should decide which mechanism is operative for each mineral. The leaching rate may sometimes be severely limited by inhibitors. These inhibitors could be organic substances or ions such as $[PO_4]^{3-}$, which deactivate the active sites on a surface (e.g. such as the effect of $[PO_4]^{3-}$ on calcite dissolution). A protective coating may sometimes also form on the surface of the weathering mineral. All these factors add to the kinetic stability of a mineral.

At this point, there are very few data on leaching rates of relevant minerals, let alone an understanding of their mechanisms. This gap certainly needs to be filled. The theoretical framework to understand the kinetics of leaching or dissolution is developed to a reasonable degree;[9,10] however, we need to apply it to relevant geologic materials.

## Crystal Chemical Criteria

### Element Substitution

In establishing which minerals are appropriate to contain the relevant nuclear waste elements, one may use minerals which are <u>known</u> to contain the element or elements of interest and which satisfy the stability criteria. Many such examples will be identified, particularly for Sr, Ln and U. However, elements such as Cs, I, An and Tc are so rare in nature that few known minerals contain them as essential elements. However, one can use the principles of crystal chemistry to predict the formation of mineral-like phases which will contain the elements in question or mineral phases into which significant quantities may be incorporated in solid solution.

The critical elements all behave essentially as ions in their compounds, so one can use the principles of element substitution in ionic compounds as criteria for predicting appropriate host phases. The main criteria are similarity of chemical parameters, particularly the ionic radius and the charge. Other parameters such as polarizability and d-orbital interactions will have a lesser effect in determining the amount of substitution. Thus one can use a table of ionic radii to predict possible substitutions, remembering that charge balance must be maintained by a coupled substitution of another element whenever necessary.

### Ionic Radii

Table P.6 lists the ionic radii of the important nuclear waste elements and of the elements present in minerals which are most likely to be substituted. Usually, complete substitution may occur if the ionic radii differ by no more than 15%. Limited substitution may occur if the radii difference is larger, or a new compound may be induced to form. This compound may be isostructural with the host phase or may have a distinctly different structure. If the phase is isostructural, then stability properties of the new phase may be similar to that of the host, certainly close enough to warrant further investigation.

Using Table P.6 as a guide, one can see that $Cs^{+1}$ is large and most like $K^{+1}$ and possibly $Ba^{+2}$. There is only one mineral in which Cs is essential, and that is pollucite $(Cs_2Al_2Si_4O_{12} \cdot nH_2O)$, a member of the analcime $(Na_2Al_2Si_4O_{12} \cdot nH_2O)$ family of minerals. The fact that it is acting in the role of $Na^{+1}$ suggests that other $Na^{+1}$ and $K^{+1}$ phases may act as hosts for $Cs^{+1}$. Other possible examples include the feldspars $(K,Na,Ca)(Al,Si)_4O_8$, feldspathoids, $(K,Na,Ca)(Al,Si_2)_3O_{4-6}$, zeolite, $(K,Na,Ca)(Al,Si)_mO_{2m} \cdot nH_2O$ and micas, $(K,Na,Ca)_2(Al,Mg,Fe)_{4-6}(Al,Si)_8O_{20}(OH)_4$. Traces of cesium are known to occur in each of these minerals.

The next element $Sr^{+2}$ is found in many compounds in nature. Often it shows substitutional relations with $Ba^{+2}$ and sometimes with $Ca^{+2}$. It may also occur in many of the

### TABLE P.6.  Selected Ionic Radii[a]

| Ion | CN[b] | Ionic Radius (Å) | Ion | CN | Ionic Radius (Å) |
|---|---|---|---|---|---|
| $Cs^+$ | X | 1.81 | $Na^{1+}$ | VI | 1.02 |
| $Sr^{2+}$ | VIII | 1.25 | | IX | 1.32 |
| $I^{1-}$ | VI | 2.20 | $K^{1+}$ | VI | 1.38 |
| $I^{5+}$ | VI | 0.95 | | IX | 1.55 |
| $Tc^{4+}$ | VI | 0.65 | $Ca^{2+}$ | VI | 1.00 |
| $Tc^{7+}$ | VI | 0.56 | | VIII | 1.12 |
| $La^{3+}$ | VIII | 1.16 | $Ba^{2+}$ | VI | 1.36 |
| ↓ | | | | VIII | 1.42 |
| $Dy^{3+}$ | VIII | 1.03 | $Cl^{1-}$ | VI | 1.81 |
| $Ce^{4+}$ | VIII | 0.97 | $Br^{1-}$ | VI | 1.96 |
| $U^{4+}$ | VIII | 1.00 | $Y^{3+}$ | VIII | 1.02 |
| $U^{6+}$ | II | 0.45 | $Zr^{4+}$ | VIII | 0.84 |
| $Np^{4+}$ | VIII | 0.98 | $Ti^{4+}$ | VI | 0.61 |
| $Pu^{3+}$ | VI | 1.00 | $Th^{4+}$ | VIII | 1.04 |
| $Pu^{4+}$ | VIII | 0.96 | $Mn^{3+}(HS)$[c] | VI | 0.65 |
| $Am^{3+}$ | VI | 1.00 | $Fe^{3+}(HS)$ | VI | 0.65 |
| $Am^{4+}$ | VIII | 0.95 | $Cr^{3+}$ | VI | 0.62 |
| $Cm^{3+}$ | VI | 0.98 | $Ce^{3+}$ | VIII | 1.11 |

a. After Shannon and Prewitt[11]

b. CN = coordination number

c. HS = high spin

same phases as indicated for $Cs^{+1}$ above.

Iodine exists in nature both as $I^-$ in two compounds and as $IO_3^-$ in several other phases. Crystal chemically as a halogen it behaves most similarly to $Br^-$ and possibly $Cl^-$, although the radii are markedly different. Very few synthetic iodine compounds have bromine or chlorine isostructural counterparts. Ways to tie it up in the crystalline state are discussed later.

Technetium is chemically most similar to manganese and rhenium. There are no known technetium compounds in nature, and there is little knowledge of its crystal chemistry. It will be discussed separately below.

The rare earth elements are all very similar in ionic size, although the heavier ones are small enough to cause them to form different series of compounds in some instances from the larger ones. For example, the large Ln's behave similarly to $Ce^{+3}$ and commonly substitute for it. The smaller Ln's tend to substitute for $Y^{+3}$. Rare earths are also known to substitute for $Th^{+4}$ and $Zr^{+4}$ in many of their minerals.

The actinides show some similarities in size and commonly follow $Y^{+3}$, $Th^{+4}$, $Zr^{+4}$, $U^{+4}$ and $Ce^{+4}$. There are enough differences between uranium chemistry and actinide chemistry to make casual geochemical reasoning suspect and specific research would be needed. Uranium readily oxidizes in nature and is commonly found as $U^{+6}$ uranates and as uranyl, $UO_2^{+2}$. Plutonyl and Neptonyl can be made and may substitute for uranyl.

**Crystalline Solutions.** Because of the ease of substitution of ions for other similar ions, it is common for solid solutions to occur. A solid solution is a compound in the crystalline state in which one or more ions have replaced other similar ions in the crystal structure without disrupting the atomic arrangement. Substitutions may be complete between two end member compositions, e.g. Fe-Mg in olivine $(Mg,Fe)_2SiO_4$, or limited, e.g. K-Na in nepheline $(Na,K)AlSiO_4$.

Natural compounds are rarely pure end members, as solid solution is very common in minerals. Some minerals may have several substitutions possible with concomitant extreme variability in chemical compositions. The amphibole family which has four different sites which allow substitution is an extreme example. Partial solid solution may actually be desirable as waste element fixation mechanism, because the mineral's stability may be better controlled by the host composition. In other words, the waste ion would be sufficiently dilute in the host structure that it does not substantially modify the stability of that host.

**Isostructural Compounds.** Crystals which allow solid solution necessarily have the same crystal structure for the end members. Compounds may exist which have the same structure which show no or very limited solid solubility, usually because of marked size differences of the ions involved. Such isostructural groups may have similar stability properties. Thus it may be useful to identify families of compounds with certain structural properties which may predict the existence of a stable compound of a particular waste element. Calcium compounds, for example, may be indicators of possible strontium compounds. Bromides and chloride compounds may be possible indicators of possible iodides. Several isostructural possibilities will be identified below.

## Synthesis

Preparation of synthetic minerals requires that the desired elements from the waste streams be mixed with other materials. The mixture is then reacted to form the synthetic mineral. Considered here are the problems that may arise in the processing of nuclear wastes into synthetic minerals.

The purity of the partitioned waste stream will determine whether side reactions will lead to additional phases in the synthetic mineral assemblage. The controlling factors will be the ionic size and the ionic charge of the additional cations present. Ions of similar size and charge to the element being packaged will dissolve into the synthetic mineral as a minor solid solution. Many of the mineral phases are very "forgiving." That is, they will accept many elements into solid solution at least in small amounts. If there is a large size or charge mismatch, the impurity elements in the waste stream will react to form secondary minerals of their own. Whether this is detrimental to the processing would have to be evaluated in individual cases.

Three general methods of reaction are in common use among geochemists for the synthesis of minerals. These may be labeled calcination, solid state reaction, and hydrothermal reaction. In each method, it is necessary to mix the waste elements with the other components in the right proportions to form the minerals. Many minerals are nearly stoichiometric, that is the components must be mixed in exactly the proportions called for in the mineral formula. If this is not done some components will be left over to form additional phases. Minerals that form solid solutions are not quite so critical.

Mineral synthesis by calcination involves these steps:

(i) Taking each component into solution (for example, as the nitrates);

(ii) Mixing the solutions in correct proportions using volumetric methods;

(iii) Precipitation of the solution as a gel, spray drying, or another method of forming a calcine (a highly reactive fine-grained, often poorly-crystallized powder);

(iv) Firing the calcine, at temperatures of typically 900 to 1400°C (temperatures depend on the mineral being synthesized) to form the final well-crystallized mineral phase.

The first step would not be necessary here since the form of the partitioned wastes is nitrate solutions. Calcination can be carried out using the types of spray calciners that have already undergone considerable development and testing for the solidification of radioactive wastes. No new technology is involved to adapt these devices to synthetic minerals and the expected difficulties are those of remote handling and metering of the solutions and of calciner operation. Firing of the calcine to form the final crystalline product in general would require temperatures that can be reached in base metal furnaces or gas-fired kilns.

Mineral synthesis by direct solid state reactions is done as the name implies. The radioactive waste and the other components needed to construct the mineral phase are mixed as solids. The solid must be intimately mixed, ground, and compacted before reaction. Reaction temperatures are higher and reaction times are longer because the components are crystalline solids and transport can only take place by diffusion. The main difficulty expected here is the maintenance of equipment at the higher firing temperatures. There would be more problems with furnace burn-out and breakage or fluxing of refractories. This is merely a special case of the general rule that experimental difficulties increase as the temperature increases. Rare earth and actinide oxides, for example, tend to be very refractory and would require high reaction temperatures if this method were employed.

Hydrothermal synthesis is the technique of reacting materials by using high pressure, high temperature water as both a solvent and as a catalyst. It has the tremendous advantage of causing reaction between poorly reactive substances at modest temperatures (200 to 800°C is about the experimental range) but it has the important difficulty of requiring reaction at high pressure, hundreds of thousands of atmospheres. Hydrothermal synthesis, although an indispensible tool in the geochemist's bag of tricks, is not suited to large scale processing. About the only commercial process that uses hydrothermal synthesis on a large scale is the growth of quartz crystals for the electronics industry. It is a batch process and inherent limitations of pressure vessels requires that the batches be fairly small. Commercial quartz-growth vessels are 2 to 3 meters high and 0.3 to 0.5 meter in diameter. To this must be added the difficulties associated with assembling and disassembling the pressure vessel by remote handling.

## DISCUSSION OF MINERAL GROUPS

### Silicate Minerals

Silica, $SiO_2$, comprises over 60% of the earth's crust, and alumina, $Al_2O_3$, another 15%. It is not surprising that these elements dominate the rock-forming minerals. About half of the known mineral species are alumino-silicates, most of which are composed of one or more of the other eleven most abundant elements in the earth's crust. Feldspar alone makes up 58% of the earth's crust. Because of the abundance of these silicate minerals and their occurrence in a wide variety of rocks, one naturally asks if any of them might be potential radionuclide hosts. Detailed chemical and crystallographic data on most of the silicate minerals have been compiled by Deer, Howie, and Zussman.[12]

The suitability of silicates as hosts depends specifically on the ability of the radionuclide to substitute in solid solution for one of the essential ions of the compound. This is especially true for the common rock-forming silicates. We will examine each of the major groups of silicate minerals considering the general principles of crystal chemistry which might elucidate any ionic substitutions of interest. We will also consider some common families of silicate minerals that appear to have potential as repository minerals.

We can dismiss some groups quite easily. The silica ($SiO_2$) family of minerals is usually rigidly stoichiometric although substitutions of Al for Si create a charge imbalance which is usually compensated for by "stuffing" the framework with $Na^+$, $K^+$ or $Ca^+$. $Cs^+$ and $Sr^{+2}$ are too large to enter into these compounds. The olivine-related minerals including the humite series are structurally based on closest packing of oxygen ions and the largest ion which seems to find its way into these compounds is $(Ca^{+2})^{VI}$ at 1.00Å. Only $Tc^{+4}$ is small enough to fit comfortably, but it is too highly charged. The Ln and An elements likewise are too highly charged.

### Pyroxene Minerals

The pyroxene group of minerals are a series of compounds with a general formula $XY(Si,Al)_2O_6$ where X represents usually a mono- or divalent ion with ionic radius in the range 0.6 to 1.0Å. Examples are $Na^+$, $Ca^{+2}$, $Mn^{+2}$, $Fe^{+2}$, $Mg^{+2}$ and $Li^+$. The Y cations are di- or trivalent ions with radii in the range 0.5 to 0.8. Examples include $Mn^{+2}$, $Fe^{+2}$, $Mg^{+2}$, $Fe^{+3}$, $Al^{+3}$, $Cr^{+3}$, and $Ti^{+4}$. These small ranges in ionic size result from a structure which is still quite closely packed in terms of the oxygen ions. Too much distortion from substitution of larger ions usually breaks down the structure.

About the only critical element which might substitute in pyroxene would be $Tc^{+4}$ with an ionic radius of 0.6Å. The only other 4-valent ion that occurs in pyroxenes is $Ti^{+4}$ (radius-- 0.605Å). Titanium rarely substitutes in quantitites greater than one percent by weight although in some of the titanaugites it may reach 3 to 5%.

The suitability of pyroxene as a technetium host would require considerable research and, as a host, pyroxenes are marginal. It is probable that ferrite-like phases will prove more suitable hosts for technetium than any silicate.

The reported rare earth content of any pyroxene is never greater than trace quantities, and these are probably due to minute inclusions of other rare earth minerals.

Pyroxenes form easily in both dry and hydrothermal systems, and they are common reaction products in many silicate experiments. In studies on the decomposition of nuclear waste products in glass under mild hydrothermal conditions pyroxene was a common end product phase. Even with the presence of all the radionuclides at moderate concentration levels, none of them was detected in the pyroxene phase.

## Amphibole Minerals

The amphibole minerals comprise a group whose general formula is $W_{0-1}X_2Y_5(Si_2Al)_8O_{22}(OH)_2$. The X and Y sites are essentially identical with those so labeled in the pyroxene minerals. The limits on ionic substitutions will be the same also. The W site which is not always occupied in amphiboles will accept low charge cations in the ionic radius range $0.95_x$Å to $1.35_x$Å. These are usually only $Na^+$ and $K^+$, and no other ions are known as substitutes. Amphiboles have sometimes been called "nature's waste-baskets" because the W, X and Y sites can accept so many elements, but the structures are not suitable for any of the critical radionuclides except possibly $Tc^{+4}$. The remarks concerning $Tc^{+4}$ are the same as for the pyroxenes discussed above.

The synthesis of amphiboles is not favorable for them to be considered as potential repository phases. Because the minerals are hydrous, water pressures must be maintained during the synthesis. This, in turn, requires that hydrothermal methods be used. Volcanic rocks rarely contain amphiboles because the water leaves the lava when it reaches the surface. Amphiboles which survive are usually formed in the magma chamber before eruption.

## Epidote Minerals

The compositional formula for the epidote minerals is $X_2Y_3Z_3(O,OH,F)_{13}$ in which

$$X = Ca, Ce^{3+}, La^{3+}, Y^{3+}, Th, Fe^{2+}, Mn^{2+}, Mn^{3+};$$
$$Y = Al, Fe^{3+}, Mn^{3+}, Fe^{2+}, Ti;$$
$$Z = Si, Be.$$

The compositions of epidote minerals which occur commonly are:

| | |
|---|---|
| zoisite/clinozoisite | $Ca_2Al_3Si_3O_{12}(OH)$; |
| epidote | $Ca_2FeAl_2Si_3O_{12}(OH)$; |
| piemonite | $Ca_2(Mn,Fe,Al)_3Si_3O_{12}(OH)$; |
| allanite | $(Ca,Ce,La,Y)_2(Mn,Fe^{2+},Fe^{3+},Al)_3Si_3O_{12}(OH)$. |

Allanite is resistant to weathering and appears as a detrital mineral.

The large X-cation site in epidote is suitable for incorporating $^{90}Sr$ as well as rare earths and possibly actinides in synthetic analogs of allanite. However, epidote would probably not be suitable as a nuclear waste host because of the difficulty in synthesizing the mineral. All of the epidote minerals are stable at low temperatures and modest to high pressure. At high temperature, greater than 600-700°C, the epidotes dissociate according to the reaction

$$4Ca_2Al_3Si_3O_{12}(OH) \rightarrow 5CaAl_2Si_2O_8 + CaSiO_3 + Ca_2Al_2SiO_7 + 2H_2O.$$
$$\text{zoisite} \qquad \text{anorthite} \quad \text{wollastonite} \quad \text{gehlenite}$$

Epidote appears readily on a laboratory time scale only at pressures in excess of 3 kilobars and temperatures in the range of 600°C.[12] Successful synthesis at atmospheric pressure by calcination or related techniques does not appear likely.

## Garnet Minerals

The garnets are orthosilicates with the general formula

$$X_3Y_2Si_3O_{12}$$

where X = Mg, $Fe^{2+}$ or Ca; Y = Al, $Fe^{3+}$ or $Cr^{3+}$.

Although the garnets are dense and close-packed structures the 8-coordinated X-cation site will accept large ions and Sr-substituted grossular ($Ca_3Al_2Si_3O_{12}$) would be of interest. However, grossular is best synthesized at temperatures in the range of 800°C under hydrothermal conditions with a water pressure of 2 kilobars. Attempts at lower pressure synthesis lead to a hydro-garnet in which OH is substituted for the oxygen or to mixtures of calcium silicates. In general, garnets are high pressure phases in nature where they occur in metamorphic rocks. Once formed the garnets are resistant to weathering and appear as detrital minerals.

## Calcium Silicate Minerals

Possible candidates among the calcium silicate minerals are limited, due partly to the hydraulic nature of the anhydrous di- and tri-calcium silicates and partly to the poor resistance of the hydrated phases to mechanical degradation and their high reactivity under quite mild hydrothermal conditions. Moreover, as with the pyroxenes to which they are related, the structures of possibly useful calcium silicate phases tend to be close-packed, with but limited possibilities for isomorphous replacement or crystalline solution, at least in the pure phases. Wollastonite ($CaSiO_3$) and rankinite ($Ca_3Si_2O_7$) appear the only serious contenders in the group. Both form from the oxides at 1200°C and represent the end-members of dehydration for hydrated calcium silicate phases. They show little reactivity at lower temperatures; in particular, neither is hydraulic. Strontium can replace calcium in both, making them possible hosts for that cation.

Possibly of more potential use are compounds closely related to the calcium silicates but with off-stoichiometric compositions. Bustamite [$(Ca,Mn,Fe)SiO_3$] and rhodonite [$(Mn,Cu)SiO_3$], formally allied to wollastonite, have more "open" structures than wollastonite and may be capable of accommodating a larger range of foreign ions in substitution. Synthesis and stability of these phases are similar to wollastonite.

Although the pure di-calcium silicates must be ruled out, appreciable amounts of lanthanide solution occurs and stabilizes the non-hydraulic, $\gamma-Ca_2SiO_4$ form. It is possible that this phase may act as a strontium and a lanthanide host, but studies are needed to define solubility limits and the stability of the material.

Recently, Scott[13] has described the crystal structure of a hydrated potassium-calcium silicate, miserite [$KCa_5(Si_8O_{22})(OH)F$], which appears capable of incorporating a wide variety of cations into a vacant site and "locking" them there. The mineral occurs with aegirine and orthoclase, sometimes with wollastonite, and appears geologically stable, and would seem a potentially useful host for a wide range of cations were some way to incorporate them into the structure to be found. Studies of the synthesis and stability of miserite could prove fruitful.

## Layer Silicate Minerals

The layer silicate minerals include the micas, the clays and the chlorite families. The mica family has the general formula $W_{0-1}Y_{2-3}(Si_2Al)_4O_{10}(OH)_2$ where W and Y have the same meaning as in the pyroxene and amphibole discussion. The same range of ionic substitutions occurs as in the amphiboles and pyroxenes. Fluorine and less commonly $Cl^-$ and $S^{2-}$ may substitute for the (OH). Biotite is commonly reported from granites and pegmatites which contain traces of rare earth elements, but these traces can usually be attributed to included xenotime $(Y...)PO_4$ inclusions rather than being incorporated into the mica structure directly.

The remarks also pertain to the other groups of layer silicates as far as ionic substitutions are concerned. Because chlorites and clays may have layer units with residual electronic charges, some ions may be adsorbed on the surfaces. Interlayer ions may be easily exchanged. The permanence of these attachments, however, is poor and the materials cannot be considered potential repository phases.

## The Melilite Minerals

The common melilites are a solid solution

$$Ca_2MgSi_2O_7 - Ca_2Al_2SiO_7$$
$$\text{åkermanite} \quad \text{gehlenite}$$

in which magnesium is gradually replaced by aluminum. The entire series can be prepared synthetically by dry-firing--that is, calcination techniques at temperatures in the range of 1000 to 1200°C. The minerals are found in nature in high temperature, low pressure environments and synthetically in slags are related materials. They appear to be stable under ambient conditions. The strontium analogs can be made and this mineral series is a potential host for $^{90}Sr$.

## Feldspar Minerals

The feldspar minerals are the most abundant mineral group on the earth and a major constituent of granite rocks but they are remarkably simple in chemistry. They have the formula $(K,Na,Ca,Ba)_1Al_{1-2}Si_{2-3}O_8$ with almost no other chemical substitutions allowed. Boron and $Fe^{+3}$ are known to substitute for Al, and Cs and Sr may substitute for the cation. A $SrAl_2Si_2O_8$ phase can be synthesized, which is analogous to $BaAl_2Si_2O_8$, but the level of Sr in natural feldspars is rarely 0.5 w/o. The level of Cs is never greater than 0.005 w/o. Feldspars weather slowly to clay minerals under surface ambients but are very stable in rocks.

## Feldspathoid Minerals

The feldspathoid minerals are those phases which form from alkali-rich aluminosilicate compositions with insufficient $SiO_2$ to form free quartz. The minerals usually coexist with feldspar, particularly the one with the corresponding alkali ion. The important feldspathoids are nepheline, $(Na,K)_4Al_4Si_4O_{16}$; leucite, $KAlSi_2O_6$; analcime, $NaAlSi_2O_6 \cdot H_2O$; sodalite, $Na_8Al_6Si_6O_{24}Cl_2$, and cancrinite, $(Na,K,Ca)_{6-8}(Al,Si)_{12}O_{24}(Cl,SO_4,CO_3)_{1.5-2.0} \cdot nH_2O$. Scapolite $(Na,Ca,K)_4Al_3(Al,Si)_3Si_6O_{24}(Cl,SO_4,CO_3)$, may also be considered here due to its similarity in behavior to sodalite and cancrinite, although it is not formally considered a feldspathoid.

Nepheline is a stuffed derivative of tridymite ($SiO_2$) and can accept alkali ions in the framework to charge compensate the Al that substitutes for Si. The cages are just large enough to accept K (ionic radius = 1.38Å) and actually prefer some Na (ionic radius--1.02Å) to relieve some of the strains on the framework linkages. To accept larger cations such as Cs and Sr would be too much strain on the structure. Cs and Sr are generally not reported in any nepheline analyses.

Leucite and analcime have similar crystal structures with identical frameworks. The cages are larger than in nepheline and Cs will substitute freely in the analcime to form the only Cs mineral in nature. Pollucite, $CsAlSi_2O_6 \cdot 0.5H_2O$, forms readily from its components, and is the leading candidate as a repository phase for Cs.[14] In fact considerable study has already been made on pollucite for this purpose. The possibility of a Sr analog also exists, but it does not occur in nature.

Sodalite, cancrinite and scapolite may play two roles as potential waste minerals although considerable research is needed to verify their potential. All three minerals may have Cs and Sr analogs with these elements substituting for Na, Ca, or K, as in leucite-analcime. The framework cages are larger than in leucite and analcime, but because of this increased size the alkali cations are easily exchanged and hence easily leachable. The other aspect of these structures which is of considerable interest is the trapping of large anions in the cages. All three minerals are known to have significant quantities of $Cl^-$, $SO_4^=$ and $CO_3^=$ in the structural cages, and sodalite often has $S_2^=$. This behavior immediately suggests the possibility of trapping $I^-$ inside the cages. If the structure can be grown around the $I^-$ before the iodine volatilizes, it may be effectively caged because its radius (2.20Å) is considerably larger than the cage opening (1.40Å). Much research is needed on this potential.

### Zeolite Minerals

The zeolites are a large group of industrially important compounds, many of which exist as minerals. Their properties have been surveyed by Breck.[15] They have aluminosilicate framework structures with larger cages and cage openings than do the feldspathoids, and all zeolites show exchange properties of the non-framework cations. This property is undesirable in a repository compound unless the radionuclide can be stabilized in the structure by effectively blocking its path to freedom.

Both Cs and Sr zeolites have been synthesized, and one Sr-zeolite occurs in nature, the mineral brewsterite, $SrAl_2Si_6O_{16} \cdot 5H_2O$. It is found in volcanic basalts in gas cavities as a very late-formed mineral.

Zeolites can be synthesized by gel and by hydrothermal methods. They contain considerable water which helps keep the framework open and which can be driven off by heat. Some structures collapse at relatively low temperatures, even as low as 100°C; but many retain their structural integrity as high as 800°C. The exchangeability of the cation, however, suggests that the zeolites in general will be ineffective in retaining desired cations for sufficient time under a variety of conditions to be effective repository compounds.

Rare earths have been exchanged in some of the zeolite phases. In particular the faujasite series may be synthesized with a Ce:Ca ratio of 6:4.[16] The faujasites are one of the more open zeolite framework structures. Considerable research is needed to determine the suitability of zeolite structures as waste repositories as they cannot be dismissed summarily.

### Borosilicate Minerals

Because boron forms a very stable oxyanion, both as $BO_3$ and $BO_4$ coordination polyhedra, many borosilicates prove to be quite stable mineral structures. Beryllium as $BeO_4$ coordination polyhedra also forms quite stable minerals with silicates. Many minerals of this type are known to contain rare earth elements either as essential elements or in solid solution to significant levels. Table P.7 lists the most important of these minerals. These minerals are considered possible repository phases.

#### TABLE P.7.

#### Borosilicate Minerals

| | |
|---|---|
| Cappelenite | $(Ba,Ca,Na)(Y,La)_6B_6Si_{13}(O,OH)_{27}$ |
| Danburite | $CaB_2Si_2O_8$ |
| Hellandite | $(Ca,Y)_2(Si,B,Al)_3O_8 \cdot H_2O$ |
| Melanocerite | $(Ce,Ca)_5(Si,B)_3O_{12}(OH,F) \cdot nH_2O$ |
| Stillwellite | $(Ce,La,Ca)BSiO_5$ |
| Tadzhikite | $Ca_3(Ce,Y)_2(Ti,Al,Fe)B_4Si_4O_{22}$ |
| Tourmaline | $(Na,Ca)(Mg,Fe...)_3Al_6(BO_3)_3(Si_6O_{18})(OH,F)_4$ |
| Tritomite | $(Ce,La,Y,Th)_5(Si,B)_3(O,OH,F)_{13}$ |
| Tinzenite | $(Ca,Mn,Fe)_3Al_2BSi_4O_{15}(OH)$ |

#### Berylosilicate Minerals

| | |
|---|---|
| Aminoffite | $Ca_2(Be,Al)Si_2O_7(OH) \cdot H_2O$ |
| Gadolinite | $Be_2Y_2FeSi_2O_{10}$ |
| Semenovite | $(Ca,Ce,La)_{12}(Be,Si)_8Si_{12}O_{40}(O,OH,F)_8 \cdot H_2O$ |
| Tugtupite | $Na_4AlBeSi_4O_{12}Cl$ |

The borosilicates and berylosilicates are primarily found in rare-earth bearing pegmatites, both granite and nepheline syenite types. The affinity for rare-earth elements is indicated by their formation. The big unknown in these phases is their stability under repository conditions. Considerable experimentation is needed to determine their suitability.

### Zirconosilicate and Titanosilicate Minerals

Interest in the zirconosilicate and titanosilicate minerals arises from the known substitution of rare-earth elements and actinides for both Ti and Zr. Usually, the quantities are small. The known minerals are listed in Table P.8. Both the zirconosilicates and titanosilicates are formed in pegmatite deposits and they are commonly associated with other rare-earth bearing minerals. Evidence suggests that many of them may be quite resistant to weathering and zircon and titanite are known to survive as heavy minerals in placer deposits.

## TABLE P.8.

### Zirconosilicate Minerals

| | |
|---|---|
| Armstrongite | $CaZrSi_6O_{15} \cdot 2.5H_2O$ |
| Bazirite | $BaZrSi_3O_9$ |
| Catapleiite | $Na_2ZrSi_3O_9 \cdot 2H_2O$ |
| Elpidite | $Na_2ZrSi_6O_{15} \cdot 3H_2O$ |
| Eudialyte | $Na_4(Ca,Ce,Fe)_2ZrSi_6O_{17}(OH,Cl)_2$ |
| Hilairite | $Na_2ZrSi_3O_9 \cdot 3H_2O$ |
| Lavenite | $(Na,Ca)_3ZrSi_2O_7(O,OH,F)_2$ |
| Lemoynite | $(Na,Ca)_3Zr_2Si_{10}O_{26} \cdot 8H_2O$ |
| Vlasovite | $Na_2ZrSi_4O_{11}$ |
| Wadeite | $K_2ZrSi_3O_9$ |
| Zircon | $ZrSiO_4$ |

### Titanosilicates

| | |
|---|---|
| Batisite | $Na_2BaTi_2Si_4O_{14}$ |
| Chevkinite | $(Ca,Ce,Th)_4(Fe,Mg)_2(Ti,Fe)_3Si_4O_{22}$ |
| Ilmajokite | $(Na,Ba,Ce)_{10}Ti_5Si_{14}O_{22}(OH)_{44} \cdot nH_2O$ |
| Joaquinite | $Ba_2NaCe_2Fe(Ti,Nb)_2Si_8O_{26}(OH,F)$ |
| Karnasurtite | $(Ce,La,Th)(Ti,Nb)(Al,Fe)(Si,P)_2O_7(OH)_4 \cdot 3H_2O$ |
| Lamprophyllite | $Na_2(Sr,Ba)_2Ti_3(SiO_4)_4(OH,F)_2$ |
| Mosandrite | $(Na,Ca,Ce)_3TiSi_2O_8F$ |
| Perrierite | $(Ca,Ce,Th)_4(Mg,Fe)_2(Ti,Fe)_3Si_4O_{22}$ |
| Titanite | $CaTiSiO_5$ |
| Tranguillityite | $Fe_8(Zr,Y)_2Ti_3Si_3O_{24}$ |
| Tundrite | $Na_3(Ce,La)_4(Ti,Nb)_2(SiO_4)_2(CO_3)_3O_4(OH) \cdot 2H_2O$ |

This group of minerals deserves more study with respect to their stabilities under repository conditions. They may actually prove to accept $Cs^+$ and $Sr^{+2}$ in some of their structures for Ca, Na or Ba. One Sr phase, lamprophyllite is known.

### Rare-Earth Silicate Minerals

There are actually a large number of minerals which are essentially rare-earth silicates with or without other essential elements. These compounds must all be considered potential repository phases for both the Ln's and An's. Some of the phases have demonstrated stabilities, having formed in granites or pegmatites and then survived the sedimentary cycle to be deposited in placers. Allanite is one example which has been discussed with the epidote minerals. Thorite, huttonite and cheralite are other examples.

Most of the minerals in Table P.9 are formed in pegmatites. The families of $Ln_2Si_2O_7$ and $Ln_2SiO_5$ phases are easy to prepare synthetically. Many of them show several structural modifications, but they have high melting or decomposition temperatures. Some of the minerals such as coffinite, $USiO_4$, may be synthesized at 100°C. These minerals form in sedimentary rocks from circulating ground waters.

## TABLE P.9.

### Rare-Earth Silicate Minerals

| | |
|---|---|
| Allanite | $(Ce,Ca,Y)_2(Fe,Al_3)(SiO_4)_3(OH)$ |
| Ashcroftine | $KNaCaY_2Si_6O_{12}(OH)_{10} \cdot 4H_2O$ |
| Britholite | $(Ca,Ce)_5(SiO_4,PO_4)_3(OH,F)$ |
| Cheralite | $(Ca,Ce,Th)(P,Si)O_4$ |
| Coffinite | $U(SiO_4)_{1-x}(OH)_{4x}$ |
| Ekanite | $(Th,U)(Ca,Fe,Pb)_2Si_8O_{20}$ |
| Huttonite | $ThSiO_4$ |
| Iimorite | $Y_5(SiO_4)_3(OH)_3$ |
| Kainosite | $Ca_2(Ce,Y)_2Si_4O_{12}(CO_3) \cdot H_2O$ |
| Miserite | $K(Ca,Ce)_4Si_5O_{13}(OH)_3$ |
| Nordite | $(La,Ce)(Sr,Ca)Na_2(Na,Mn)(Zn,Mg)Si_6O_{17}$ |
| Phosinaite | $H_2Na_3(Ca,Ce)SiO_4PO_4$ |
| Sazhinite | $Na_3CeSi_6O_{15} \cdot 6H_2O$ |
| Soddyite | $(UO_2)_5Si_2O_9 \cdot 6H_2O$ |
| Thalenite | $Y_2Si_2O_7$ |
| Thorite | $ThSiO_4$ |
| Thorosteenstrupine | $(Ca,Th,Mn)_3Si_4O_{11}F \cdot 6H_2O$ |
| Thortveitite | $(Sc,Y)_2Si_2O_7$ |
| Tombarthite | $Y_4(Si,H_4)_4O_{12-x}(OH)_{4+2x}$ |
| Törnebohmite | $(Ce,La)_3Si_2O_8(OH)$ |
| Umbozerite | $Na_3Sr_4ThSi_8(O,OH)_{24}$ |
| Uranophane | $Ca(UO_2)_2(SiO_3OH)_2$ |
| Weeksite | $K_2(UO_2)_2Si_6O_{15} \cdot 4H_2O$ |
| Yttrialite | $(Y,Th)_2Si_2O_7$ |

Again, this group of minerals requires considerable research to define the suitability of any of the rare-earth silicates as repository phases. Their long-time stability must be defined particularly under hydrothermal conditions.

## Oxide Minerals

### Perovskite Structure--$ABO_3$ ($CaTiO_3$)

A = Ca, REE, Na, Th, U      ~ radius   1.0Å
B = Ti, Nb, Ta, $Fe^{3+}$, Mg, Zr    ~ radius   0.7Å
Knopite   $(Ca,Ce)(Fe,Ti)O_3$
Dysanalyte   $(Ca,Ce,Na)(Ti,Nb,Fe)O_3$
Loparite   $(Na,Ce,Ca)(Ti,Nb)O_3$
Irinite   $(Na,Ce,Th)_{1-x}(Ti,Nb)O_{3-x}(OH)_x$
Metaloparite   $(Ce,Ca)_{1-x}(Ti,Nb)_{3-x}(OH)_2$

Loparite, irinite and knopite are found as metamict minerals. Perovskite occurs as an accessory mineral in basic igneous rocks, often in association with melilite, nepheline or rare earth apatite, as well as in metamorphosed calcareous rocks in contact with basic igneous rocks. The B ion is mostly Ti with a little Nb and $Fe^{3+}$ in all the various minerals above. The variety rich in rare earths, chiefly cerium, is knopite and if also high in alkalis (Na), loparite, or its hydrate, metaloparite. Dysanalyte is high in Nb and irinite is distinguished by its high thorium content.

Since $Ca^{2+}$ is in 12-fold coordination in perovskite, it is replaced preferentially by the large light lanthanides, i.e. La, Ce ($r_{La3+}$ = 1.15Å, $r_{Ce3+}$ = 1.11Å) rather than the yttrium earths. Hydrothermal alteration of loparite leads to metaloparite with loss of alkalis, assimilation of water and enrichment in the rare earth elements.[17] Thus it seems that loparite retains the REE in alteration. Loparite is also known to occur as a placer deposit-forming mineral. Therefore, perovskite minerals are a possible host for Ln and An elements.

We can calculate the conversion of perovskite to rutile by a weathering solution, i.e.

$$CaTiO_3 + 2H^+ \rightarrow Ca^{2+} + H_2O + TiO_2$$

$$K_{298} = 10^{18.14}$$

Hence for pH = 6    $a_{Ca2+} = 10^{6.14}m$;

pH = 8    $a_{Ca2+} = 10^{2.14}m$.

Evidently the reaction, at equilibrium, proceeds overwhelmingly to the right, suggesting that loss of Ca (and maybe REE) would follow if equilibrium were maintained. However, the kinetics of the above reaction may be slow, and more work is needed to determine the leaching rate.

Pyrochlore--$A_2B_2O_6$(O,F,OH) or $(Ca,Na,Ce)_{2-x}(Nb,Ti)_2O_6(OH,F)$

The pyrochlores are also characteristic of basic rocks and alkali rock massifs (nepheline-syenites, alkali syenites, albitized granites and carbonatites) and occur in close association with albite, zircon, apatite, sphene, biotite. Pyrochlore occurs in both the metamict and crystalline state. There is quite a variety of names:

Pyrochlore $(Na,Ca,U,Ce,Y)_{2-x}(Nb,Ta,Ti)_2O_6(OH,F)$;
Betafite $(U,Ca)_{2-x}(Nb,Ti,Ta)_2O_{6-x}(OH)_{1+x}$, high Ti and U;
Zirconolite $(CaZrTi_2O_7)$;
Microlite $(Ca,Na)_2Ta_2O_6(O,OH,F)$, high Ta;
Djalmaite $(Ca,Na,U)_2Ta_2O_6(O,OH,F)$, high U relative to microlite;
Obruchenite $(Y,U,Ca)_{2-x}Nb_2O_6(OH)$, low Ti, high Y and U.

The differences among minerals reflects only the amounts of U, Ti, Ta, Y relative to pyrochlore. Pyrochlore from carbonatites can have up to 4% $ThO_2$. Hydration of pyrochlore leads to loss of mobile REE, Ca, Na and an increase in U.[17] Pyrochlore can have up to 19% $U_3O_8$ and high Sr. Pyrochlore also occurs as a placer deposit-forming mineral.

$AB_2O_6$--Nb-Ti-Ta Oxides

Columbite Structure. Columbite $(Fe,Mn)(Nb,Ta)_2O_6$ (tantalite). Columbite can have up to 3% REE, little U.

It is very abundant in acid rocks, e.g. (rarer) granite, granitic pegmatites, quartz veins; occurs in association with biotite, albite, zircon. Columbite-tantalite is a placer

deposit-mineral and is insoluble in acids.[17] Furthermore, it is very resistant to weathering and accumulates in deluvial, eluvial, and alluvial placers, resulting from the weathering
of columbite-bearing granite and pegmatite. In placers, it is associated with cassiterite,
zircon, ilmenite and rutile. Columbite may be a good candidate for hosting Ln and An elements.

> Euxenite Structure. Euxenite-polycrase $Y(Nb,Ti)_2(O,OH)_6$--$Y(Ti,Nb)_2(O,OH)_6$
> Delorenjite $Y(Ta,Nb)_2(O,OH)_6$
> Fersmite $(Ca,Ce)(Nb,Ti,Fe)_2(O,OH,F)_6$.

Y is in higher coordination than in columbite. Th, U, Ca can replace Y up to several
percent, U up to 16% $UO_2$, Th up to 8% $ThO_2$. They are widespread in granite pegmatites.
Euxenite occurs as accessory mineral in granites and is also found in small amounts in
placers. It is associated with ilmenite, monazite, xenotime, zircon, garnet.

Fersmite is found in nepheline-syenite and carbonatite massifs in association with
columbite, apatite, calcite, fluorite. It is typical of rocks of intermediate composition
(for weathering and alteration see below).

> Priorite Structure. Priorite-Aeschynite $(Ce,Nd,Th,Y)(Ti,Nb)_2O_6$
> Polymignite $(Ca,Fe,Ce)(Zr,Ti,Nb)_2O_6$
> Sinicite $(Ce,Nd,Th,U)(Ti,Nb)_2O_6$, high U.

Priorite differs from euxenite by having cerium RE's and a high content of thorium and
Zr (little U). The REE have the same coordination as in the euxenite structure. Aeschynite
occurs as an accessory in some deposits related to nepheline-syenite and alkali-syenite
massifs in association with zircon, biotite, corundum, muscovite, sphene, fluorite.

The weathering and alteration of the $AB_2O_6$ and $A_2B_2O_6(O,OH,F)$ REE-Nb-Ti-Ta complex
oxides can be handled in one group. These oxides have pervasive alteration with a usual
weathered crust surrounding fresher oxides.[18] The results of weathering are leaching of
the A-site cations (i.e. U,REE) and introduction of $H_2O$ or $OH^-$ or $O^=$ into the oxide. The B
cations remain basically unchanged.[18,19]

In weathering, up to 40% decrease in the REE content is possible, although the REE
distributions remain nearly the same.[18] For example, a priorite from the Kibara Mountains,
North Katanga, had a fresh inner zone (black) with $\sim$ 0.075 cerium atoms and 0.95 U atoms per
5.58 O atoms and one brown-yellow altered outer zone with $\sim$ 0.03 cerium atoms and 0.05 U atoms
per 4.94 O atoms. Wambeke[19] gives the relative leaching rate of A cations as 110 REE atoms,
120 Na atoms and 40 U atoms per 100 atoms of Ca leached out. There are little hard data on
the kinetics or solubility of these complex oxides and should be obtained. It seems that
columbite might be a good candidate among this group for Ce disposal, since it can be very
resistant to alteration. Euxenite is the candidate for the U elements.

> $ABO_4$ Oxides.

> Fergusonite Structure. A = Y, REE, U, Ca, Th
> B = Nb, Ta, Ti.

Solid solution: $YNbO_4$ - $YTaO_4$.
fergusonite    formanite

The REE in fergusonite are mostly the yttrium rare earths.[17] Fergusonite occurs as a meta-mict mineral. It is fairly abundant in granite pegmatites. It accumulates in small amounts in placers and is found as an accessory mineral in granites. In pegmatites, it is associated with zircon, monazite, xenotime and euxenite. A study of monazite-bearing alluvial deposits in Malaya[20] showed fergusonite occurring with columbite, Ta/Nb rutile, cassiterite and garnet. The samples were derived from a cassiterite-bearing granite. It thus seems that fergusonite might be relatively stable as a host of REE and actinides.

## Carbonate and Sulfate Minerals

### Rare Earth Fluoro-Carbonates

Carbonate minerals are compounds of some cations with the carbonate anion, $CO_3^{-2}$, often with hydroxyls and waters of hydration. Of more than 70 naturally occurring carbonate compounds, most are either water soluble or are easily decomposed. These include the simple and complex carbonates of the alkali metals, the alkaline earth metals, and the transition metals. Most carbonates are sensitive to pH and dissolve easily in low pH solutions.

Exceptions to the general instability of carbonate minerals are the fluorocarbonate compounds of the rare earths. These are:

| | |
|---|---|
| Bastnaesite | $(Ce,La)CO_3F$; |
| Parisite | $Ca(Ce,La)_2(CO_3)_3F_2$; |
| Cordylite | $Ba(Ce,La)_2(CO_3)_3F_2$; |
| Synchisite | $Ca(Ce,La)(CO_3)_2F$. |

Bastnaesite and parisite are relatively insoluble even in low pH solutions at ambient temperatures. None are insoluble in hot, low pH solutions. The rare earth fluorocarbonates could act as hosts for rare earth elements in neutral or alkaline repository rocks.

### Sulfate Minerals

The number of sulfate minerals numbers several hundred but nearly all are soluble in water or are otherwise unstable. Two exceptions of interest are barite, $BaSO_4$, and celestine, $SrSO_4$. The solubility of barite in cold water is only 2.2 ppm while the solubility of celestine is 113 ppm. There is a complete solid solution between barite and celestine although intermediate compositions are not found in nature.

Use of barite and celestine as hosts for $^{90}Sr$ would be of value in a bedded anhydrite repository (anhydrite = $CaSO_4$) because of the chemical compatibility.

## Phosphate Minerals

The phosphate-containing minerals include a subset, which seems particularly suited to the disposal of nuclear waste elements. Natural phosphate minerals are all orthophosphates, the major one being fluor-apatite. There are two major families of relevance to nuclear waste disposal: the apatite family and the monazite-xenotime family.

Since in nature phosphorus will exist in only one valence state (+5) (for example, $H_2PO_3^- > H_2PO_4^-$ only when $f_{O_2} < 10^{-101}$ at 250°C), the distribution and stability of its species in solution will be Eh-independent. On the other hand, the dominant phosphorus species in solution will be strongly dependent on pH and on possible complexing cations, since $PO_4^\equiv$, $HPO_4^=$

and $H_2PO_4^-$ form strong complexes [e.g. with uranium[21]]. The reaction

$$H_2PO_4^- \rightarrow HPO_4^= + H^+ \qquad (1)$$

has a $\Delta G_r^\circ$ = 9.83 kcal/mole and a $\Delta H_r^\circ$ = +0.99 kcal/mole at 25°C, which yields a $K_1$ = $10^{-7.21}$ at 25°C. Hence, for pH < 7.21, $H_2PO_4^-$ will be the dominant $PO_4$ species in solution and for pH > 7.21, $HPO_4^=$ will be dominant. Ignoring complexes, this will also be true at higher temperatures, since $\Delta H_r^\circ$ is so small. The total phosphorus content of ground waters, $\Sigma PO_4$, is most often greater than 0.1 ppm but rarely greater than 1 ppm, so

$$0.1 \text{ ppm} < \Sigma PO_4 < 1 \text{ ppm}.$$

In the mineral structure, the $PO_4$ tetrahedra can often be replaced by the $CO_3$, $SO_4$ and $SiO_4$ groups, leading to a variety of phosphate minerals.

### Apatite Family - $Ca_5(PO_4)_3(OH,F)$

Apatite is the most abundant phosphorus-bearing mineral. It is a common accessory mineral in many types of rocks (acid to basic). Apatite can take up significant amounts of Sr (up to 11.6 wt % SrO) and also rare earths (up to 11 wt % REE) and so may be a suitable host for nuclear waste elements. The rare earths, predominantly Ce, may replace Ca in apatites of alkaline igneous rocks. $U^{+4}$ (r = 0.97Å) can also substitute for $Ca^{2+}$ (r = 0.99Å) and natural apatites have ~0.01% U, if primary igneous apatite, or slightly richer, 0.02% U if sedimentary marine apatite. Th is more abundant than U by a factor of 3 or 4.[12] Apatites can contain $CO_3$, $SO_4$ and $SiO_4$ groups replacing $PO_4$. In sedimentary phosphorites, the apatite can have up to 7-8% $CO_3^=$ content, with much lesser $SO_4$ or $SiO_4$ substitution. The carbonate content of on-shore phosphorites is less (3%) than that of sea floor phosphorites, suggesting that weathering reduces the carbonate content.

In terms of geologic evidence for stability to weathering, apatite is not uncommon in sedimentary rocks where it occurs both as a detrital mineral and as a primary deposit. It is not classified as a placer deposit-forming mineral, however. On the weathering stability list of Pettijohn[6], apatite has an index of 6, putting it beneath biotite and garnet. Smithson[22] from a study of Jurassic sandstones in Yorkshire, England, lists apatite as stable in unweathered rock but decomposed in weathered rocks. Graham[23] lists apatite with olivine as least stable and Jackson[5] puts it low in the second stage of the weathering sequence of clay-size mineral particles. Thus the stability of apatite has yet to be firmly shown.

Strontium apatite results in the solid solution:

$$Ca_5(PO_4)_3F \text{ -NaCeSr}_3(PO_4)_3(OH).$$

$$\text{fluor-apatite} \qquad \text{belovite}$$

However, belovite is unstable under surface conditions and is readily replaced by rhabdophan-ite, $CePO_4 \cdot H_2O$, leading to a rapid loss of Sr and Na.[17] There is unlimited substitution in the systems $Ca_5(PO_4)_3F-Sr_5(PO_4)_3F$ and $Ca_5(PO_4)_3(OH)-Sr_5(PO_4)_3(OH)$. Sr-apatite, found in alkali pegmatites, is readily soluble in acids.[17]

We can use the solubility criteria laid out in the introduction. Although thermodynamic data for Sr-apatite are lacking, there are data for fluor- and hydroxy-apatite.[24] Using these data, we can compute the following:

$$Ca_5(PO_4)_3F + 3H^+ \rightarrow 5Ca^{2+} + 3HPO_4^= + F^-$$
$$K_1 = 10^{-33.33} \times 10^{1997.7[1/T-1/298]}$$
(2)

$$Ca_5(PO_4)_3F + 6H^+ \rightarrow 5Ca^{2+} + 3H_2PO_4^- + F^-$$
$$K_2 = 10^{-11.70} \times 10^{2646.8[1/T-1/298]}$$
(3)

$$Ca(PO_4)_3(OH) + 4H^+ \rightarrow 5Ca^{2+} + 3HPO_4^= + H_2O$$
$$K_3 = 10^{-12.17} \times 10^{7051.0[1/T-1/298]}$$
(4)

$$Ca_5(PO_4)_3(OH) + 7H^+ \rightarrow 5Ca^{2+} + 3H_2PO_4^- + H_2O$$
$$K_4 = 10^{+9.46} \times 10^{7698.0[1/T-1/298]}$$
(5)

If we use $\Sigma PO_4 = 10^{-6}$m ($\sim$0.1 ppm) and $a_{F^-} = 1.6 \times 10^{-5}$m ($\sim$0.3 ppm), typical values for ground waters, we obtain the following values for the activity of calcium in equilibrium with the apatites.

Fluor-apatite

$a_{Ca^{2+}}$

| pH/T | 25°C | 75°C |
|------|------|------|
| 6 | $1.04 \times 10^{-5}$m | $5.81 \times 10^{-6}$m |
| 8 | $1.24 \times 10^{-7}$m | $7.95 \times 10^{-8}$m |

Hydroxy-apatite

$a_{Ca^{2+}}$

| pH/T | 25°C | 75°C |
|------|------|------|
| 6 | $1.23 \times 10^{-3}$m | $2.24 \times 10^{-4}$m |
| 8 | $5.83 \times 10^{-6}$m | $1.22 \times 10^{-6}$m |

In ground waters, $a_{Ca^{2+}}$ is typically $\sim 10^{-3}$m.[8] Therefore in alkaline environments we expect both apatites to be stable at all temperatures 25°C--100°C. However, in acid environments hydroxy-apatite will not be stable, while fluor-apatite will be somewhat stable, more so at higher temperatures. Chien[25] has also shown that the carbonate substitution may _increase_ the equilibrium dissolution of apatite.

Monazite-Xenotime (Ce,La)PO_4-YPO_4

This family is one of the most promising for the disposal of nuclear wastes. Both monazite (Ce,La)PO_4, and xenotime, YPO_4, as well as their hydrates, rhabdophanite, (Ce,Ca)PO_4·H_2O, and churchite, YPO_4·H_2O, are simple orthophosphates. They are always crystalline even though they may contain significant amounts of U and Th. Monazite is isostructural with huttonite, ThSiO_4, and xenotime with zircon (ZrSiO_4) and coffinite (USiO_4). Monazite can contain quite a high content of thorium (28%) by the substitution $Th^{4+} + Si^{4+} \rightarrow Ce^{3+} + P^{5+}$ (i.e. ThSiO_4-CePO_4 solid solution). Monazite is a selective cerium mineral (i.e. high radius rare earths). It has lesser amounts of uranium (up to 4%).[12,17] It is sparingly soluble in acids and is very stable under weathering conditions, often collecting in placers formed from the disintegration of monazite-containing granites. It occurs as an accessory in granites and granitic pegmatites and is abundant in metamorphic deposits.[17] It occurs as a detrital mineral in sands from weathering of granites and gneisses.

Dryden and Dryden[26] compared the changes in relative abundance of various minerals from the fresh rocks to the weathered products in samples from the Wissahickon schist in Pennsylvania and Maryland. They found, by taking the ratios of the number of grains of each mineral in fresh and weathered rock, that the resistance of zircon relative to garnet is 100 (i.e. garnet/$Zr_{fresh}$/garnet/$Zr_{weathered}$ ~100), sillimanite 40, monazite 40, chloritoid 20, kyanite 7 and all other minerals less than 5. This is in agreement with Pettijohn[6] who ranked monazite in his "weathering sequence" as 3 after zircon (1) and tourmaline (2). The general geologic evidence points to a very resistant mineral.

We can calculate the solubilities for monazite to establish its thermodynamic stability. Taking $\Sigma PO_4 = 10^{-6}$m (0.1 ppm), we can compute the solubility of $Ce^{3+}$ in a natural leaching solution as a function of pH, and temperature. The thermodynamic data for $CePO_4$ were obtained from Naumov, et al.[24] We obtain:

$$CePO_4 + 2H^+ \rightarrow Ce^{3+}(aq) + H_2PO_4^-$$

$$\Delta G_r^\circ = 3.27 \text{ kcal/mole} \qquad \Delta H_r^\circ = -11.71 \text{ kcal/mole} \tag{6}$$

$$CePO_4 + H^+ \rightarrow Ce^{3+}(aq) + HPO_4^=$$

$$\Delta G_r^\circ = 12.10 \text{ kcal/mole} \qquad \Delta H_r^\circ = -10.72 \text{ kcal/mole.} \tag{7}$$

Therefore

$$K_6 = 4.00 \times 10^{-3} \, e^{5893.3[1/T - 1/298]}$$

$$K_7 = 2.46 \times 10^{-10} \, e^{5395.1[1/T - 1/298]}.$$

Assuming no complexing, pure solids, and $\Sigma PO_4 = 10^{-6}$m, then

$$a_{Ce^{3+}}(aq)$$

| pH/T | 25°C | 50°C |
|------|------|------|
| 6 | $4.0 \times 10^{-9}$m | $8.6 \times 10^{-10}$m |
| 8 | $2.46 \times 10^{-12}$m | $6.1 \times 10^{-13}$m |

The low values of $a_{Ce^{3+}}$ obtained support the stability evidence from the geologic data. Obviously monazite is more stable in warm alkaline environments. Increasing the phosphate content of the ground water would also further stabilize the monazite. Thus if $\Sigma PO_4 = 10^{-5}$m (1 ppm), $a_{Ce^{3+}} = 4.0 \times 10^{-10}$m at pH = 6, T = 25°C and the same for the other conditions.

Xenotime contains a high amount of yttrium and the yttrium rare earths. It is widespread in granites, pegmatites and metamorphic gneisses.[17] When granites weather, xenotime accumulates in placers (e.g. in New Zealand and USSR). Xenotime is very stable under surface conditions.

## Iodine Hosts

### Iodine Minerals

Iodine is a relatively rare element in rocks and minerals. It occurs in both the $I^-$ and $I^{5+}$ valence states. Iodine is easily oxidized to the 5-valent state and appears in many of its natural compounds as the iodate, $IO_3^-$ ion. These are:

| | |
|---|---|
| Lautarite | $Ca(IO_3)_2$; |
| Bellingerite | $Cu(IO_3)_2 \cdot 2/3H_2O$; |
| Salesite | $Cu(IO_3)OH$; |
| Schwartzembergite | $Pb_5(IO_3)Cl_3O_3$; |
| Dietzeite | $Ca_2(IO_3)_2CrO_4$. |

The above compounds are at least slightly soluble in water, and all are soluble in solutions with low pH. The iodate minerals are found in evaporite deposits or as weathering products of ores in very dry environments.

Marshite, CuI, iodargyrite, AgI, and their solid solution, miersite, occur in nature and might be stable in a bedded salt type of repository but in general it may be said that there is no natural mineral of iodine that hints of very long term stability.

### Framework Structures for Iodine

Two candidate minerals that are composed of three-dimensional frameworks that contain cavities sufficiently large to house the $I^-$ ion are sodalite and the boracite family.

Sodalite, $Na_4Al_3Si_3O_{12}Cl$, is a member of the feldspathoid group. It is a three-dimensional framework and the essential $Cl^-$ is locked in cage-like interstices. Iodine can be substituted for $Cl^{-1}$ and maintained in this structure.

Boracite, $Mg_3B_7O_{13}Cl$ is a three-dimensional framework of B-O tetrahedra with the $Cl^-$ locked in a cage structure. Other minerals of the boracite family are ericaite, $(Fe,Mn)_3B_7O_{13}Cl$, and chambersite, $Mn_3B_7O_{13}Cl$. However, a very large number of synthetic materials with the boracite structure have been synthesized. Many of the synthetics contain $I^-$ rather than $Cl^-$. They are stable under hydrothermal conditions, although difficult to synthesize by ceramic methods.

### Lead Oxyhalides

There exists a small group of minerals composed of the oxy- or hydroxy-halides of lead. These materials usually appear as oxidation products on lead-zinc ores which is evidence for their stability in the surface environment. The list includes

| | |
|---|---|
| Murdochite | $PbCu_6(O,Cl,Br)_8$; |
| Mendipite | $Pb_3Cl_2O_2$; |
| Penfieldite | $Pb_2Cl_3(OH)$; |
| Yedlinite | $Pb_6CrCl_6(O,OH)_8$; |
| Phosgenite | $Pb_2(CO_3)Cl_2$. |

Little is known of these minerals. The substitution of iodine for chloride in the lead oxy-halide structures should be investigated. Little is known of the structures, solubilities, and ranges of stability of these minerals, although it is doubtful that they would have great thermal stability.

## Uranium Minerals

Uranium occurs in nature in both the $U^{+4}$ and $U^{+6}$ valence state. The $U^{+5}$ valence state has been postulated especially in $U_3O_8$ and other oxides intermediate between $UO_2$ and $UO_3$, but it has not really been verified. Its existence is not critical to the discussion.

### $U^{+4}$ Minerals

There is only a small number of minerals in which uranium occurs as $U^{+4}$. The most important and best known is uraninite, $UO_2$, which has the fluorite, $CaF_2$, structure. It is the principal mineral in most uranium deposits and is found in pegmatites, in sandstones and metasediments, and as an accessory mineral in some granites. Natural $UO_2$ is rarely stoichiometric and is better described as $UO_{2+x}$ where x ranges between 0 and 0.25. Most uraninite from older sources is metamict and may be called pitchblende.

In sandstone deposits the uraninite has formed from circulating ground water by reduction of the $U^{+6}$. In the reduced form it is very stable and is known in the placer deposits of the Witwatersrand district in Africa as a common mineral. These uraninite grains were carried down streams and deposited in energetic depositional environments without chemical breakdown because the atmospheric conditions of the time were highly reducing. If uraninite could be maintained in its $U^{+4}$ state it would be a good repository mineral. Unfortunately, it alters rapidly in present-day atmospheres.

Uraninite is usually only uranium bearing in sandstone deposits, but in pegmatites it may contain significant quantitites of Ce and Th in solid solution. Actually, complete solid solutions of these elements can be prepared under laboratory conditions.

Some of the other $U^{+4}$ minerals occur in quantities sufficient for them to be called ore minerals. Coffinite, $USiO_4$; brannerite, $UTi_2O_6$, and ningyoite, $CaU(PO_4)_2 \cdot 1.5H_2O$, occur primarily in sedimentary or metasedimentary environments probably as syngenetic minerals. Other $U^{+4}$ minerals include lermontovite, $(U,Ca,Ce...)_3(PO_4)_4 \cdot 6H_2O$; sedovite, $U(MoO_4)_2$; uranopyrochlore, $U_2Nb_2O_6(O,OH,F)$; cliffordite, $UTe_3O_8$, and ishakowaite, $(U...)(Nb,Ta)O_4$. In addition $U^{+4}$ occurs as a minor element in many minerals mostly replacing other group IV elements or the rare earths. At the conditions existing at the earth's surface all these $U^{+4}$ minerals readily alter by oxidation and weather by releasing the uranium into the ground water system. The $U^{+6}$ may be fixed immediately in new minerals or may migrate for long distances before being redeposited.

### Uranate Minerals

Uranium as $U^{+6}$ forms a large group of oxides, hydrated oxides, and uranates. The uranates form compounds with Na, K, Mg, Ca, Ba and Pb. Some of these compounds are anhydrous, but most are hydrates. There are many crystalline modifications of $UO_3$ but none occurs naturally. Usually the hydrate schoepite, $UO_3 \cdot 2H_2O$ or one of its polymorphic forms occurs. If the other elements are present the tendency is to form the uranate minerals.

The uranates occur in the immediate vicinity of the source mineral, usually uraninite. They develop as a replacement aureole of poorly crystallized phases commonly called gummite. The Pb which is common in older deposits is primarily radiogenic in origin.

The uranates do not survive further weathering and are replaced by uranyl compounds in the main oxidized zone of any ore body. It is doubtful if any uranate would be a good uranium repository.

### Uranyl Minerals

Any uranium which finds its way into the ground water system migrates as the uranyl ion, $UO_2^{+2}$, or as some complex involving the uranyl ion. As the local conditions change the uranyl ion may precipitate as one of over 100 mineral species.

### Uranyl Ion

The uranyl ion is a linear group with the uranium in the center and the oxygen ions on the ends. Because of this unique geometry uranyl compounds form their own series of compounds in nature with very little substitution of other ions.

Uranyl will form complex structures with almost any oxyanion, carbonate, sulfate, phosphate, arsenate, molybdate, selenate, vanadate and silicate. The crystal structure of the minerals usually comprises uranyl-oxyanion sheets or chains which stack so as to contain interstitial low charge cations and water molecules. Most of the carbonates, sulfates, molybdates and selenates and even the silicates are moderately soluble and will leach as the environmental conditions change. The phosphates-arsenates and vanadates appear to be very insoluble and may be potential repository compounds. The known minerals are listed in Table P.10.

The uranyl phosphates and arsenates are usually considered together because their crystal chemistry is very similar and in some cases there is even partial substitution of phosphorus and arsenic. In all compounds these ions exist in tetrahedral coordination. In a few vanadates, the vanadium is tetrahedral, but in most vanadates it forms complex $V_2O_8$ groups of pentagonal edge-shared $VO_5$ coordination polyhedra. Thus most of the vanadates must be separated from the others.

As can be seen in Table P.10, the phosphates-arsenates-vanadates are usually classified by their U:X ratio where X is P, As, V. Several ratios exist but the most commonly found group in nature is the U:X = 1. Within this group are several minerals which have great potential as repository minerals. This potential is suggested by the wide range of occurrence, the frequency of mineral formation and the extremely low solubility of the compounds.

The abundance of uranyl phosphates and arsenates is more due to the stability of uranyl phosphate and uranyl arsenate complexes in ground water[21] rather than to any abundance of P or As. The complex polymerizes readily into sheet-like crystal structures which incorporate a variety of low charge cations and water molecules between the sheets. Thus, they form a large number of mineral species depending on the available cation.

Considering the phosphates and arsenates first and especially the phosphates, because the toxicity of As makes it an undesirable additive, the most important group is the family of autunite minerals. The family is usually broken into three groups--autunite, meta-autunite I, and meta-autunite II, depending on the number of water molecules involved. Table P.11 lists all the members of the autunite family. The variation of water is common to the group but does not seem to affect the stability of the species themselves.

TABLE P.10. Uranyl Phosphates, Arsenates, Vanadates

$UO_2:XO_4$

| 4:2 | Arsenuranylite | $Ca(UO_2)_4(AsO_4)_2(OH)_4 \cdot 6H_2O$ |
| | Bergenite | $Ba(UO_2)_4(PO_4)_2(OH)_4 \cdot 8H_2O$ |
| | Renardite | $Pb(UO_2)_4(PO_4)_2(OH)_4 \cdot 7H_2O$ |
| 3:2 | Troegerite | $(UO_2)_3(AsO_4)_2 \cdot 12H_2O$ (see 2:2) |
| | Huegelite | $Pb_2(UO_2)_3(AsO_4)_2(OH)_4 \cdot 3H_2O$ |
| | Dumontite | $Pb_2(UO_2)_3(PO_4)_2(OH)_4 \cdot 3H_2O$ |
| | Phosphuranylite | $Ca(UO_2)_3(PO_4)_2(OH)_4 \cdot 7H_2O$ |
| 2:2 | Carnotite | $K_2(UO_2)_2(V_2O_8) \cdot 3\text{-}5H_2O$ |
| | Tyuyamunite | $Ca(UO_2)_2(V_2O_8) \cdot 5\text{-}8H_2O$ |
| | Metatyuyamunite | $Ca(UO_2)_2(V_2O_8) \cdot 3H_2O$ |
| | Curienite* | $Pb(UO_2)_2(V_2O_8) \cdot 5H_2O$ |
| | Francevillite* | $Ba(UO_2)_2(V_2O_8) \cdot 5H_2O$ |
| | Strelkinite | $Na_2(UO_2)_2(V_2O_8) \cdot 6H_2O$ |
| | Autunite* | $Ca_{1\text{-}2}(UO_2)_2(PO_4)_2 \cdot 8\text{-}12H_2O$ |
| | Meta-autunite I* | $Ca_{1\text{-}2}(UO_2)_2(PO_4)_2 \cdot 6\text{-}8H_2O$ |
| | Meta-autunite II* | $Ca_{1\text{-}2}(UO_2)_2(PO_4)_2 \cdot 4\text{-}6H_2O$ |
| | Meta-vanuralite | $Al(UO_2)_2(VO_4)_2(OH) \cdot 8H_2O$ |
| | Vanuralite | $Al(UO_2)_2(VO_4)_2(OH) \cdot 11H_2O$ |
| | Vanuranylite | $(H_3O,Ba,Ca,K)_{1.6}(UO_2)_2(VO_4)_2 \cdot 4H_2O$ |
| | Dewindtite | $Pb(UO_2)_2(PO_4)_2 \cdot 3H_2O$ |
| | Sengierite | $Cu(UO_2)_2(V_2O_8) \cdot 8\text{-}10H_2O$ |
| 2:3 | Coconinoite | $Fe_2^{3+}Al_2(UO_2)_2(PO_4)_2(SO_4)(OH)_2 \cdot 20H_2O$ |
| 2:4 | Parsonsite | $Pb_4(UO_2)_2(PO_4)_2 \cdot 2H_2O$ |
| | Przhevalskite | $Pb(UO_2)_2(PO_4)_4 \cdot 4H_2O$ |
| | Pseudoautunite | $(H_3O)_4Ca_2(UO_2)_2(PO_4)_4 \cdot 5H_2O$ |
| | Walpurgite | $(BiO)_4(UO_2)_2(AsO_4)_4 \cdot 6H_2O$ |
| | Hallimondite | $Pb_2(UO_2)_2(AsO_4)_2$ |

Autunites are known to form compounds with Ca, Mg, Ba, Na, Cu, $Fe^{2+}$, K, Zn, Mn, Co, Pb, $NH_4$, Al, and $H_2O$. Many synthetic analogs can also be easily formed including Sr and even Li. The included cation is easily exchangeable in acid solutions but the autunite structure remains unaffected by the many substitutions.

In nature the specific species autunite, $Ca(UO_2)_2(PO_4)_2 \cdot 8\text{-}12H_2O$, and meta-autunite I, $Ca(UO_2)_2(PO_4)_2 \cdot 6\text{-}8H_2O$, are the most common and could in fact be considered a very common mineral anywhere uranium exists. It is found as a secondary mineral in all climates from arid to humid. It has been mined as an ore mineral in several locations because of its abundance. In Cameron, Arizona, it occurs in near-surface sandstone lenses and around Shoshoni, Wyoming it is mined from bentonite pits where it forms in the desiccation cracks of the clay. At Ningyo Prefecture in Japan it occurs in sandstone where it was mined extensively

TABLE P.11. The Autunite Family

Autunites, $R_{1-2}(UO_2)_2(XO_4)_2 \cdot 8\text{-}12H_2O$

| | | |
|---|---|---|
| Autunite | $Ca(UO_2)_2(PO_4)_2 \cdot 8\text{-}12H_2O$ | |
| Fritschelite | Mn | V |
| Heinrichite | Ba | As |
| Kahlerite | Fe | As |
| Novacekite | Mg | As |
| Sabugalite | H,Al | P |
| Saleeite | Mg | P |
| Sodium autunite | Na,Ca | P |
| Torbernite | Cu | P |
| Uranocircite | Ba | P |
| Uranospinite | Ca | As |
| Zeunerite | Cu | As |

Meta-autinites, $R_{1-2}(UO_2)_2(RO_4)_2 \cdot 6\text{-}8H_2O$

| | | |
|---|---|---|
| Abernathylite | $K_2(UO_2)_2(AsO_4)_2 \cdot 6\text{-}8H_2O$ | |
| Bassettite | $Fe^{2+}$ | P |
| Meta-ankoleite | $K_2$ | P |
| Meta-autunite I | Ca | P |
| Metaheinrichite | Ba | As |
| Metakahlerite | $Fe^{2+}$ | As |
| Metakirchleimerite | Co | As |
| Metalodevite | Zn | As |
| Metanovacekite | Mg | As |
| Metaforbernite | Cu | P |
| Meta-uranocircite | Ba | P |
| Meta-uranospinite | Ca | As |
| Metazeuerite | Cu | As |
| Sodium uranospinite | NaCa | As |
| Troegerite | $(H_3O)_2(UO_2)_2(AsO_4)_2 \cdot 6H_2O$ | |
| Uramphite | $NH_4$ | P |
| unnamed | $(H_3O)_2$ | P |
| Meta-autunite II | $Ca(UO_2)_2(PO_4)_2 \cdot 4\text{-}6H_2O$ | |

until the primary ningyoite zone was encountered. Some very noted specimen localities include the Daybreak Mine in Washington, and Cornwall, England. It is also a common alteration product in uranium-bearing pegmatites. In all these localities it has proven itself to be a very stable phase. Its leaching characteristics under various conditions still must be tested.

Among the other uranyl phosphates several other candidates are also evident as possible repositories. In particular we should consider the phosphuranylite $Ca(UO_2)_4(PO_4)_2(OH)_4 \cdot 7H_2O$. It is a much rarer mineral than autunite but has a higher loading factor because the U:P ratio is 3:2. Considerably less is known about the stability of this phase. Its conditions of formation and synthesis are less well known but it occurs similarly to autunite.

One must not overlook the vanadates as potential repository minerals, in particular carnotite, $K_2(UO_2)_2V_2O_8 \cdot 3\text{-}5H_2O$; tyuyamunite, $Ca(UO_2)_2V_2O_8 \cdot 5\text{-}8H_2O$, and metatyuyamunite, $Ca(UO_2)_2V_2O_8 \cdot 3H_2O$. These three minerals occur extensively throughout the Colorado Plateau and have been mined for uranium. They usually occur in sandstone lenses forming interstitially among the sand grains. Once formed they appear to resist weathering and alteration even at surface conditions. Strontium analogs might easily be made. Ion exchange, common in the autunites, does not seem to occur in the vanadates.

## Technetium Hosts

Since the element technetium is not known in nature, it follows that no minerals exist with technetium as an essential element. Tc exists mainly in valence states $Tc^{4+}$ and $Tc^{7+}$ with the latter forming the very soluble pertechnatate ion. $Tc^{4+}$ forms stable solid oxide phases and, because of a similar ionic radius, behaves much like $Ti^{4+}$. Many titanium analogs have been synthesized[27] including spinels, pyrochlores, perovskites, and a stable solid solution between $TiO_2$ and $TcO_2$. Titanium minerals may be the best hosts for technetium if reducing conditions are maintained in the repository.

## MINERAL TABLES

### Hosts for Radionuclides

Table P.12 lists selected minerals which have potential as hosts for radionuclides. The entries in Table P.12 were selected according to the criteria listed below.

Approximately 2500 mineral species have been identified. These have been compiled into reference sources of which those of the Dana system,[28] Deer, Howie and Zussman,[12] Strunz,[29] and Roberts, Rapp and Weber[30] were consulted. Each of the 2500 minerals was reviewed and in a first sieving all minerals that were known to be water-soluble, chemically undesirable or crystal chemically unsuitable as radionuclide hosts were eliminated. A much shortened list of about 100 minerals remained. A second sieving eliminated minerals of great chemical complexity that would be difficult to synthesize. The minerals that remained were separated according to the radionuclide for which they were to serve as host and these groups were then roughly ranked with the best candidates listed first.

Table P.13 is the final listing. In addition to mineral name and formula, the table lists some available information on the occurrence of these minerals in nature, which provides clues to their stability in the repository environment, and on alteration processes where known. It must be emphasized that the data on these later categories are very sparse although this study does not claim to be an exhaustive literature survey. Table P.13 is intended as a guide for future research rather than finalized data for engineering design.

### Commentary on Table P.12

The lack of silicate minerals on the listing is perhaps unexpected. Silicates make up the bulk of the rocks on the earth and many of them are very stable. However, the common silicate structures utilize the most abundant elements of the earth and the critical radionuclides from nuclear waste are, with the exception of $^{90}Sr$, unusual elements, either too large or too small to fit into available sites in the silicate minerals. Furthermore, silicates are relatively less resistant to weathering and only a few, zircon being an outstanding example, survive the weathering process to become detrital minerals and even fewer survive to become placer minerals.

Phosphates first, and oxides second, rise to the top of the list as the most stable minerals in a wide variety of geochemical environments from initial formation at high temperatures and pressures, through weathering, transport, contact with salt water in oceanic depositional basins, burial, diagenesis, upheaval, and in some cases a complete second cycle of weathering.

A very large number of the phases on the list occur in pegmatites or in alkaline rocks which are closely related. The minerals, by implication, are stable in the presence of aqueous solutions at temperatures and pressures to 600°C and several kilobars. Chemical compatibility with granite rocks is implied. Whether many of these minerals would be compatible with other candidate repository rocks, basalts and shales would require research. The fact that the minerals do not occur in these rocks in nature means only that the chemistry for their formation was not correct, not that the minerals are necessarily incompatible.

The rankings except for the top few entries are almost arbitrary. Although available mineralogical evidence suggests that these minerals are stable in the temperature and pressure regime generally thought to exist around nuclear waste repositories, their _relative_ stabilities are not known. Likewise, the _relative_ solubilities of these generally insoluble phases are not known. Thus, detailed ranking or the construction of any sort of figure of merit cannot be done under the present state of knowledge.

Many of the oxide minerals are highly stable and insoluble because of a particular oxidation state. Lower oxidation states of the transition metals and of uranium form less soluble compounds than do the higher oxidation states. The state of oxidation in a repository would be controlled by the oxidation potential and oxygen buffer capacity of the host rocks since these are present in vastly larger volumes than the volume of the waste. Likewise the solubilities of many of the minerals are a sensitive function of the acidity of any circulating solutions. The fluorocarbonates are an example of minerals with low solubilities in neutral or alkaline solutions but which become progressively more soluble as the pH decreases. The host rock in which the repository is formed will play an important role in buffering the oxidation potential and acidity of any circulating ground water that might contact the synthetic minerals of the waste form.

The large number of minerals that are listed as occurring in pegmatites is to be expected. Pegmatites are complex mineral assemblages which form from a residual high-water content fluid that remains after the crystallization of granitic rocks. Ions which are too big or too small or have the wrong charge or the wrong electronic structure to fit into any of the common granite minerals--quartz, feldspars, micas, and amphiboles, are concentrated in the residual fluid and finally crystallize into pegmatites. It is not the pegmatite-forming temperature and pressure regime that is critical but rather the complex solution chemistry that allows these minerals to be formed. Many of these minerals can be synthesized by entirely different methods but their occurrence in pegmatites does imply a substantial degree of mutual compatibility among the phases.

TABLE P.12. Selected Host Minerals for Radionuclides

| Element | Host Mineral | Formula | * | Occurrence in Nature | Alteration |
|---------|--------------|---------|---|---------------------|-----------|
| Cs | pollucite | $Cs_{2-x}Na_xAl_2Si_4O_{12} \cdot H_2O$ | E | granite pegmatites | |
| Sr | anorthite (feldspar) | $Ca_{1-x}Na_xAl_{2-x}Si_{2+x}O_8$ | R | basalts | slow breakdown into clay minerals under surface weathering conditions |
| | Sr-apatite | $Sr_5(PO_4)_3(OH,F)$ | E | alkalic pegmatites | |
| | belovite | $(Sr,Ce,Na,Ca)_5(PO_4)_3(O,OH)$ | SS | alkalic pegmatites | breakdown at low pH |
| | celestine | $SrSO_4$ | E | oxidation zones in sulfur deposits primary precipitation | |
| | Sr-autunite | $Sr(UO_2)_2(PO_4)_2$ | E | strata-bound ore deposits | |
| | goyazite | $SrAl_3(PO_4)_2(OH)_5H_2O$ | E | pegmatite | |
| | lamprophyllite | $Na_2(Sr,Ba)_2Ti_3(SiO_4)_4(OH,F)_2$ | SS | nepheline syenites alkali-rich pegmatites | |
| | lusangite | $(Sr,Pb)Fe_3(PO_4)_2(OH)_5 \cdot H_2O$ | SS | pegmatite | |
| | bøggildite | $Na_2Sr_2Al_2PO_4F_9$ | E | cryolite deposits | |
| | danburite | $CaB_2Si_2O_8$ | R | andesite xenoliths | |
| | attakolite | $(Ca,Mn,Sr)_3Al_6(PO_4,SO_4)_7 \cdot 3H_2O$ | SS | | |
| | cuspidine | $Ca_4Si_2O_7(F,OH)_2$ | R | metamorphic rocks limestone contact zones | |
| | rankinite | $Ca_3Si_2O_7$ | R | strain zones | gelatinizes readily at low pH |
| | melilite | $Ca_2Mg_{1-x}Al_{2x}Si_{2-x}O_7$ | R | extrusive rocks | |
| | umbozerite | $Na_3Sr_4ThSi_8(O,OH)_{24}$ | E | | |
| | scheelite | $CaWO_4$ | R | pegmatites | |
| | powellite | $CaMoO_4$ | R | pegmatites | |
| I | sodalite | $Na_8Al_6Si_6O_{24}Cl_2$ | R | nepheline-syenite rocks | |
| | boracite | $Mg_3B_7O_{13}Cl$ | R | salt domes and salt deposits | occurs in the "water insoluble" fractions of salt deposits |
| | ericaite | $(Fe,Mn)_3B_7O_{13}Cl$ | R | salt domes and salt deposits | occurs in the "water insoluble" fractions of salt deposits |
| | chambersite | $Mn_3B_7O_{13}Cl$ | R | salt domes and salt deposits | occurs in the "water insoluble" fractions of salt deposits |

TABLE P.12. Selected Host Minerals for Radionuclides (continued)

| Element | Host Mineral | Formula | * | Occurrence in Nature | Alteration |
|---------|-------------|---------|---|---------------------|------------|
| I | parahilgardite | $Ca_2B_5O_8Cl(OH)_2$ | R | salt domes | occurs in "water insoluble" fraction |
| | murdochite | $PbCu_6(O,Cl,Br)_8$ | R | oxidation zones of Pb-Zn deposits | |
| | mendipite | $Pb_3Cl_2O_2$ | R | | |
| | penfieldite | $Pb_2Cl_3(OH)$ | R | | |
| | yedlinite | $Pb_6CrCl_6(O,OH)_8$ | R | | |
| | phosgenite | $Pb_2(CO_3)Cl_2$ | R | | |
| | marshite | $CuI$ | E | associated with copper ores | darkens on exposure to air |
| | iodargyrite | $AgI$ | E | secondary mineral in silver ores | |
| | miersite | $(Ag,Cu)I$ | E | associated with copper ores | |
| Tc | perovskite | $CaTiO_3$ | R | basic igneous rocks | |
| | calzirtite | $CaZr_3TiO_9$ | R | carbonatite | partially dissolves in low pH solutions |
| | yttrocrasite | $(Y,Th,U,Ca)_2Ti_4O_{11}$ | R | | |
| | batisite | $Na_2BaTi_2Si_4O_{14}$ | R | nepheline syenite | |
| | brannerite | $(U,Ca,Ce)(Ti,Fe)_2O_6$ | R | hydrothermal mineral | |
| Ln,An | monazite | $(Ce,La)PO_4$ | E ⎫ | granites, pegmatites, placers, hydrothermal deposits metamorphic rocks | extremely stable |
| | cheralite | $(Ce,Ca,Th)(P,Si)O_4$ | SS ⎭ | | sometimes yellow crust of rhabdophanite |
| | xenotime | $YPO_4$ | E | granites, pegmatites, placers, hydrothermal deposits, sandstones | very stable alters to churchite |
| | rhabdophanite | $(Ce,La)PO_4 \cdot H_2O$ | E ⎫ | alkali pegmatites, hydrothermal deposits, sandstones | very stable forms from monazite but dehydrates to monazite on prolonged storage |
| | brockite | $(Ca,Th,Ce)PO_4 \cdot H_2O$ | E ⎬ | | |
| | grayite | $(Th,Pb,Ca)PO_4 \cdot H_2O$ | E ⎭ | | |

TABLE P.12. Selected Host Minerals for Radionuclides (continued)

| Element | Host Mineral | Formula | * | Occurrence in Nature | Alteration |
|---|---|---|---|---|---|
| Ln,An | churchite | $YPO_4 \cdot 2H_2O$ | E | alkali massifs limonite ores | forms from xenotime |
| | zircon | $ZrSiO_4$ | R | acid and alkali igneous rocks pegmatites, placers | metamict highly resistant to weathering |
| | baddeleyite | $ZrO_2$ | R | carbonatites, gabbro, placers, basalts | highly stable |
| | tacheranite | $(Zr,Ca,Ti)O_2$ | R | alkali massifs | |
| | bazirite | $BaZrSi_3O_9$ | R | granites | |
| | zirkelite | $Zr(Ca,Th,Ce)(Ti,Nb)_2O_7$ | SS | magnetite deposits pyroxenites | |
| | thorite | $ThSiO_4$ | E | greisens from granites | metamict |
| | huttonite | $ThSiO_4$ | E | sands | |
| | thalenite | $Y_2Si_2O_7$ | E } | pegmagites | alters to Y-bastnaesite |
| | yttrialite | $(Y,Th)_2Si_2O_7$ | E } | | |
| | thortveitite | $(Sc,Y)_2Si_2O_7$ | E | pegmatites | |
| | bastnaesite | $(Ce,La)CO_3F$ | E | hydrothermal deposits, pegmatites, granites | gradual alteration to lanthanite, rhadophanite or cerianite |
| | cordylite | $Ba(Ce,La)_2(CO_3)_3F_2$ | E | alkali syenite | |
| | parisite | $Ce_2Ca(CO_3)_3F_2$ | E | detrital, hydrothermal deposits, pegmatites, carbonate ore bodies | replaced by bastnaesite |
| | synchysite | $CaCe(CO_3)_2F$ | E | alkali syenite pegmatite | |
| | röntgenite | $Ce_3Ca_2(CO_3)_5F_3$ | E | pegmatite | |
| | cerianite | $(Ce,Th)O_2$ | E | carbonates, pegmatites | |
| | davidite | $(Fe,La,Ce,U)_2(Ti,Fe)_5O_{12}$ | SS | granites, skarns, pegmatites, with vein minerals | |
| | euxenite | $Y(Nb,Ti)_2(O,OH)_6$ | E } | pegmatites, granites, placers | can be altered but somewhat stable |
| | polycrase | $Y(Ti,Nb)_2(O,OH)_6$ | E } | | |
| | delorenzite | $Y(Ta,Nb)_2(O,OH)_6$ | E } | | |
| | fersmite | $(Ca,Ce)(Nb,Ti,Fe)_2(O,OH,F)_6$ | SS } | | |
| | columbite | $(Fe,Mn)(Nb,Ta)_2O_6$ | R } | granites, pegmatites, quartz veins, greisen deposits, placers | very resistant to weathering |
| | tantalite | $(Fe,Mn)(Ta,Nb)_2O_6$ | R } | | |

TABLE P.12.  Selected Host Minerals for Radionuclides (continued)

| Element | Host Mineral | Formula | * | Occurrence in Nature | Alteration |
|---------|-------------|---------|---|---------------------|------------|
| Ln,An | perovskite | $CaTiO_3$ | R | basic igneous rocks | can be altered to metaloparite but retains lanthanides |
| | loparite | $(Na,Ce,Ca)(Ti,Nb)O_3$ | SS | alkali syenites | |
| | aeschynite | $(Ce,Nd,Th,Y)(Ti,Nb)_2O_6$ | SS | | |
| | polymignyte | $(Ca,Fe,Ce)(Zr,Ti,Nb)_2O_6$ | SS | alkali massifs pegmatites | usually weathers |
| | sinicite | $(Ce,Nd,Th,U)(Ti,Nb)_2O_6$ | SS | | |
| | fergusonite | $YNbO_4$ | E | granitoid formations, placers, granites, pegmatites | fairly stable often associated with monazite in placers |
| | formanite | $YTaO_4$ | E | | |
| | samarskite | $(Fe,Y,U)(Nb,Ti,Ta)_2O_7$ | SS | pegmatites, gold placers | |
| | pyrochlore | $(Na,Ca,U,Ce,Y)_{2-x}(Nb,Ta,Ti)_2O_{6-x}(OH,F)_{1+x}$ | SS | | |
| | betafite | $(U,Ca)_{2-x}(Nb,Ti,Ta)_2O_{6-x}(OH)_{1+x}$ | SS | | |
| | zirconolite | $CaZrTi_2O_7$ | R | alkali rock massifs | pervasively altered |
| | microlite | $(Ca,Na)_2Ta_2O_6(O,OH,F)$ | R | | |
| | obruchevite | $(Y,U,Ca)_{2-x}Nb_2O_6(OH)$ | SS | | |
| | djalmaite | $(Ca,Na,U)_2Ta_2O_6(O,OH,F)$ | SS | | |
| | pandaite | $(Ba,Sr)_{2-x}(Nb,Ti)_2O_{7-x} \cdot 7H_2O$ | R | | |
| U | uraninite | $UO_2$ | E | pegmatites | rapid in oxidizing conditions but very stable in reducing conditions |
| | carnotite | $K_2(UO_2)_2(VO_4)_2$ | E | sandstone | relatively insoluble |
| | tyuyamumite | $Ca(UO_2)_2(VO_4)_2$ | E | | |
| | autunite | $Ca(UO_2)_2(PO_4)_2$ | E | pegmatites, sandstone, sedimentary breccia | insoluble |
| | K-autunite | $K_2(UO_2)_2(PO_4)_2$ | E | | |
| | Sr-autunite | $Sr(UO_2)_2(PO_4)_2$ | E | | |
| | phosphuranylite | $Ca(UO_2)_4(PO_4)_2(OH)_4 \cdot 7H_2O$ | E | U-schists, pegmatites | |
| | ningyoite | $(U,Ca,Ce)_2(PO_4)_2 \cdot 1-2H_2O$ | E | sedimentary rocks | |
| | lermontovite | $(U,Ca,Ce)_3(PO_4)_4 \cdot 6H_2O$ | E | | |
| | coffinite | $U(SiO_4)_{1-x}(OH)_{4x}$ | · E | sandstone, sedimentary breccia, U-schists | |
| | ekanite | $(Th,U)(Ca,Fe,Pb)_2Si_8O_{20}$ | SS | pegmatite veins | |

TABLE P.12.  Selected Host Minerals for Radionuclides (continued)

| Element | Host Mineral | Formula | * | Occurrence in Nature | Alteration |
|---------|--------------|---------|---|----------------------|------------|
| U | weeksite | $K_2(UO_2)_2Si_6O_{15}\cdot 4H_2O$ | E | | |
| | soddyite | $(UO_2)_5Si_2O_9\cdot 6H_2O$ | E | pegmatite | |

---

*substitution of radionuclide into host mineral:  E = essential element; SS = solid solution; R = replacement of another element by radionuclide.

## METAMICTIZATION

Metamict minerals are a special class of amorphous materials which were initially crystalline.[31]  Although the mechanism for the transition is not clearly understood, radiation damage caused by alpha particles and recoil nuclei is certainly critical to the process.[32,33] The study of metamictization of naturally occurring materials allows for the evaluation of the long-term effects that result from this type of radiation damage, particularly changes in physical properties.  Comparison of metamict and non-metamict crystalline phases addresses the question of the susceptibility of different bonding and structure types to radiation damage and provides useful insights into defining radiation damage experiments.

### Properties

The list below is an amplified tabulation of properties listed by Pabst.[34]

1. They are generally optically isotropic but may show varying degrees of anisotropy.  Reconstitution of birefringence with heating is common.

2. Metamict phases lack cleavage.  Conchoidal fracture is characteristic.

3. Some mineral species are pyronomic, that is, they glow incandescently on heating.  In many cases, however, recrystallization may occur without observable glowing.

4. Crystalline structure is reconstituted by heating.  The metamict material recrystallizes to a polycrystalline aggregate with a concomitant increased resistance to attack by acid.  During recrystallization several phases may form, the particular phase assemblage is dependent on the conditions of recrystallization (e.g., temperature and type of atmosphere). In many cases the original pre-metamict phase may not recrystallize due to compositional changes caused by post-metamict alteration.

5. Metamict minerals contain U and Th, although contents may be quite variable (as low as 0.41% $ThO_2$ in gadolinite from Ytterby, Norway).  Rare earth elements are also common (in some cases over 50 weight percent).  Water of hydration may be high (up to 70 mole percent).

6. They are x-ray amorphous.  Partially crystalline metamict minerals display distinct line broadening and decreased line intensities.  A shift of lines to lower values of two-theta is observed in specimens with a reduced specific gravity.

7. Some phases occur in both the crystalline and metamict state, and in these cases there is little chemical difference.

The most common methods of analysis of the metamict state are x-ray diffraction analysis of annealed material[35-38] and differential thermal analysis (DTA).[39-41] Most of the effort by mineralogists has been directed at establishing identification criteria.

Elemental analysis is commonly completed by wet chemical means on mineral separates or by standard electron microprobe analysis. The presence of water, both structural and absorbed, and the preponderance of rare earth elements make a complete chemical analysis a rarity in the literature.

Although radiation damage experiments are voluminous, there have been only limited and unsuccessful efforts to simulate the process of metamictization under laboratory conditions.[42,43]

## Summary of Observed Metamict Phases

In order to understand the compositional and structural controls on the process of metamictization, it is useful to tabulate naturally occurring phases which are known to exist in the metamict state. Table P.13 lists those phases which were described as being partially or completely metamict in a review of the literature by Bouska.[44] This tabulation lists only the major compositional end-members. As one might expect for mineral groups of complex compositions (e.g., compare the A:B ratios for fergusonite and samarskite) which are metamict and much altered, the nomenclature of any single mineral group is quite complicated and much confused by the proliferation of varietal names.[45] For a more detailed listing and discussion of the mineralogical literature the reader is referred to Bouska.[44]

The asterisk by each mineral name indicates it also occurs as a partially or completely crystalline phase. In some cases (e.g., monazite, xenotime and vesuvianite) the inclusion of a mineral phase as metamict is based only on a single or poorly documented occurrence. In these instances the critical reference is indicated. In other cases (e.g., rutile) the radiation damage was not due to constituent uranium and thorium nuclides but rather occurred only along grain boundaries where the rutile was in epitaxial contact with radioactive davidite.

For those phases which occur in both crystalline and metamict forms, it is interesting to compare their uranium and thorium contents. Table P.14 gives the average $U_3O_8$ and $ThO_2$ contents of orthorhombic $AB_2O_6$-type Nb-Ta-Ti oxides. Although the data in the literature are limited, in general those specimens of euxenite, fersmite, aeschynite and lyndochite which are found in the crystalline state have distinctly lower uranium and thorium contents than their metamict euxenite and aeschynite counterparts. A similar relation has been demonstrated for zircons.[46,47]

Table P.15 is a compilation of radioactive minerals which are said to be always crystalline. Comparisons of Tables P.13 and P.14 quickly reveals inconsistencies which are present in the literature. Huttonite is listed as always crystalline[34] and partially metamict.[44] Many of these inconsistencies may be resolved by very detailed and specific examinations of nomenclature. Also, note that among the phases listed as metamict (e.g., columbite and stibiotantalite), their structures probably will not accommodate either uranium or thorium. Reports of radioactive columbites are almost certainly mixtures of columbite and metamict microlite.[36] A number of the phases (bastnaesite and all hydrated phases) are alteration

### TABLE P.13. Systematic Tabulations of Metamict Minerals[a]

SIMPLE OXIDES
  *Uraninite ($UO_2$)[b,c]
  *Rutile ($TiO_2$)

PHOSPHATES
  *Monazite[48,49]
  *Xenotime[53]
  *Griphite[54]

SILICATES
  Nesosilicates (Si:O = 1:4)

    *Zircon
    *Thorite
    *Coffinite
    *Titanite[51]
    *Huttonite
    *Steenstrupine-Cerite
    *Britholite group
      *Lessingite
      Karnasurtite
      Karnocerite
      Tritomite
      Spencite
      Rowlandite
      Gadolinite

  Sorosilicates (Si:O = 2:7)

    Thortveitite group
      *Thalenite
      Yttrialite
      *Hellandite
      *Rincolite
      Epidote group
        *Allanite
        *Chevkinite
        *Perrierite
      *Vesuvianite[44]

  Cyclosilicates (Si:O = 1.3)

    *Eudialyte
    Cappelenite[52]

Nb-Ta-Ti OXIDES (A = U, $Th_1$, REE, Co, Na, K, Mg, Mn, $Fe^{+2}$, Pb; B = Nb, Ta, Ti, $Fe^{+3}$, W)

  $ABO_3$ (Perovskite structure)
    *Loparite
    Irinite
    *Knopite

  $A_{2-x}B_2O_{7-3}\cdot nH_2O$ (Pyrochlore structure)
    *Pyrochlore
    Betafite
    *Microlite
    Djalmaite
    Obruchevite
    *Zirconolite

  $A_2B_5O_{15}$ (Davidite structure)
    *Davidite

  $ABO_4$ (Fergusonite structure)
    Formanite
    *Fergusonite
    Risorite

  $AB_2O_6$ (Columbite structure)
    *Columbite[50,45]

  $AB_2O_6$ (Euxenite structure)
    *Euxenite[55]
    Polycrase
    Delorenzite
    *Fersmite

  $AB_2O_6$ (Priorite structure)
    *Priorite
    *Aeschynite
    *Bloomstrandine
    Polymignite

  $AB_2O_4$ (Samarskite structure)
    Samarskite
    Chlopinite
    Loranskite
    Yttrocrasite

  $AB_2O_6$ (Brannerite structure)
    *Brannerite
    Thorutile

  $AB_2O_7$ (Zirkelite structure)
    Zirkelite

---

a. After Bouska, reference 44.

b. The (*) indicates that the mineral also occurs as a partially or completely crystalline phase.

c. A reference indicates that inclusion of the mineral in this table is based only on a single or poorly documented occurrence.

TABLE P.14. Uranium and Thorium Content (wt %) of Non-Metamict and Metamict $AB_2O_6$--Type Nb-Ta-Ti Oxides

|  | $U_3O_8$ | $ThO_2$ |
|---|---|---|
| Non-metamict | | |
| euxenite [56] | a | a |
| fersmite [57] | a | a |
| aeschynite [58,59] | not detected | 0.72 |
| allanite [60] | 0.25[b] | 2.26[b] |
| lyndochite [61] | 0.08[c] | 3.75 |
| Metamict | | |
| euxenites (mean value of 28 analyses) | 9.31 | 3.08 |
| aeschynites (mean value of 22 analyses) | 1.2 | 10.73 |

a. semiquantitative analysis, no U or Th reported

b. analysis by R. C. Ewing, University of New Mexico

c. reported as $UO_3$

TABLE P.15. Radioactive Minerals Reported as Always Crystalline [a]

| | |
|---|---|
| autunite | $Ca(UO_2)_2(PO_4)_2 \cdot 10-12H_2O$ |
| bastnaesite | $(Ce,La)(CO_3)F$ |
| carnotite | $K_2(UO_2)_2(VO_4)_2 \cdot 3H_2O$ |
| columbite | $(Fe,Mn)(Nb,Ta)_2O_6$ |
| gummite | $UO_3 \cdot nH_2O$ |
| *huttonite | $ThSiO_4$ |
| metatorbernite | $Cu(UO_2)_2(PO_4)_2 \cdot 8H_2O$ |
| *monazite | $(Ce,Th)PO_4$ |
| stibiotantalite | $Sb(Ta,Nb)O_4$ |
| *thorianite | $ThO_2$ |
| thortveitite | $(Sc,Y)_2Si_2O_7$ |
| tyuyamunite | $Ca(UO_2)_2(VO_4)_2 \cdot nH_2O$ |
| uvanite | $U_2V_6O_{21} \cdot 15H_2O$ |
| *xenotime | $(Y,U)PO_4$ |
| yttrofluorite | $Ca_3YF_9$ |
| titanite | $CaTiSiO_5$ |
| *uraninite | $UO_2$ |
| *baddeleyite | $ZrO_2$ |

*Primary phases which are invariably crystalline, even with high concentrations of uranium and thorium. Note that in some rare cases even these minerals have been reported as being partially metamict.

a. After Ueda, reference 62.

products. The primary phases which consistently occur in crystalline form, even with high concentrations of uranium or thorium, are indicated by asterisks.

## Rate of Metamictization

The rate of metamictization of a given mineral is, to a first approximation, dependent on: (1) the inherent stability of its structure and (2) the alpha particle flux resulting from the presence of uranium, thorium and their unstable daughter nuclides.[34]

Pabst calculated that a minimum of 110,000 years is required for gadolinite, 0.4% Th, to become metamict. This figure, which could be low by a factor of 1000,[62-64] was obtained by assuming that all of that alpha decay energy was spent in disordering the structure and that this energy was measurable by DTA.[34]

Most zircons become metamict upon receiving a radiation dose of about $10^{16}$ $\alpha$/mg.[46] Using this dosage criterion, the following table gives estimates of the time required for some radioactive zircons to become metamict.

| Initial radionuclide content | Estimated time (yrs) |
|---|---|
| 1% Th | $14 \times 10^8$ |
| 1% U | $3.3 \times 10^8$ |
| 10% U | $0.32 \times 10^8$ |
| 1% $Pu^{236}$ (does not exist in nature) | 2.0 |

There are, however, zircons and thorites (thorite has the zircon structure and is expected to show similar radiation effects) which show anomalous radiation effects. Some zircons which have had radiation doses of only $2.8 \times 10^{15}$ $\alpha$/mg are metamict.[47] On the opposite extreme is a report of a non-metamict thorite containing 10% uranium that is at least $120 \times 10^6$ years old.[65] If this age is correct, then the thorite specimen has withstood a radiation dose of about $9 \times 10^{16}$ $\alpha$/mg. These data suggest that factors other than structural stability and alpha particle flux are important in determining the rate of metamictization.

## Alteration Effects

Minerals that occur in the metamict state are often severely altered, either as a result of hydrothermal alteration or surface weathering. The resulting complicated compositional variations are in part responsible for the very complex mineral nomenclature. Most of the available data on alteration effects pertains to various Nb-Ta-Ti oxides[18,19] and zircon, $(Zr,U)SiO_4$. In both cases alteration may be extensive and followed by recrystallization of phases quite different from the original pre-metamict phase.[66]

For metamict, $AB_2O_6$-type, Nb-Ta-Ti oxides (A = REE, $Fe^{2+}$, Mn, Ca, Th, U, Pb; B = Nb, Ta, Ti, $Fe^{3+}$) primary hydrothermal alteration causes a consistent increase in calcium content, generally a decrease in the uranium and thorium content, a decrease in total rare earth concentrations, a slight decrease in B-site cations, and an increase in structural and absorbed water. Secondary alteration caused by weathering is similar in effect but produces a decrease in Ca content, an increased leaching of A-site cations and a relative increase in B-site cations. Refractive index, specific gravity and reflectance decrease with both types of alteration, but $VHN_{50}$ remains approximately constant. It is important to note that although alteration effects in these natural materials have been carefully documented,

there are no experimental data on hydrothermal alteration effects, solubility as a function
of degree of metamictization, or the kinetics of these reactions.

There is an abundant literature on metamictization and alteration effects observed
in zircon, $(Zr,U)SiO_4$, a phase commonly used by geologists in U/Pb radiometric dating.
A summary of this literature is beyond the scope of this Appendix, but it should be
the subject of future research.  Discordant ages reported for metamict zircons indicate
that the U/PB rations can be changed or slightly disturbed by alteration.[67]  Laboratory
experiments involving zircon have demonstrated that altered regions are more rapidly
dissolved by 48% hydrofluoric acid.  There are some data which suggest that zircons
that have become metamict are susceptible to attack by solutions that can cause alteration. [67,68]  However, Mumpton and Roy [69] have recrystallized numerous metamict
zircons by hydrothermal treatment at temperatures of 500°C and above, and found that the
Zr:Si ratio remained close to 1:1.  This is an indication that neither element was
selectively dissolved.  They also demonstrated that the water often found in metamict
zircons was molecular $H_2O$ and not the result of $H^+$ ion exchange leaching.  The data are
still too limited to draw broad conclusions regarding the effect of metamictization
on solubility, even for metamict zircons.  Yet, at worst, this does not seem to be a
major problem.  It should be noted that monazite, a mineral that apparently does not
metamictize, was chosen as the (Ln + An) synthetic mineral in the reference scenario.

SOME NEW "SYNTHETIC MINERAL" CANDIDATES

## SOME NEW "SYNTHETIC MINERAL" CANDIDATES

Three important candidate structures for containing fission product radionuclides which do not either occur as natural minerals at all, or <u>do not contain the ions of interest in their natural analogues,</u> are important enough to be singled out for attention: hollandite; magnetoplumbite; and NZP (sodium zirconium phosphate). These phases all have considerable leach resistance in the laboratory although we have no mineral weathering data on the compositions of interest. Our detailed knowledge of these families comes <u>not</u> from mineralogy but from ceramic science. They have in common the interesting property that the cations inhabit relatively open, large channels. Hence each was first considered as an important candidate for ultra-fast cationic conductors, and only were recetnly for radionuclide immobilization.

### Hollandite

This natural mineral has the approximate composition $(Ba,K_2)O \cdot Mn_2O_3 \cdot 6(Mn^{4+}O_2)$. This phase is pseudo-tetragonal, with $a_o$ = 9.90Å and $c_0$ = 2.86Å. Considerable work has been done on this structure with the composition approximating to $BaAl_2Ti_6O_{16}$ in the search for fast-ion conductors. Ringwood first proposed a cesium-containing version of this as a radwaste ceramic phase.[70] The precise composition in normal (simulated) waste assemblages appears never to have been determined, but recently it has been reported that the Cs-pure material with composition $CsTi_6O_{13}$ has been synthesized.

### Magnetoplumbite

This is a very rare mineral of the composition $(Pb^{2+}Mn^{2+})O \cdot 6(Fe^{3+},Mn^{3+},Ti^{4+})_2O_3$. In technology the structure with several minor variants appears in different fields. It is very widely produced as the principal magnetically-hard ferrite $BaO \cdot 6Fe_2O_3$. More recently another variant has attracted ever greater attention as the main ceramic material for fast-ion transport of $Na^+$ in the compound $Na_2O \cdot 11Al_2O_3$ or β-alumina. Its structure consists of interleaved blocks having a spinel-like structure, and layer units with large cation spacings. It has the important crystal chemical property of accommodating large ions of any charge. Thus $Cs^+$ can easily substitute for the $Na^+$; or $Ca^{2+}$, $Sr^{2+}$, $Ba^{2+}$ for two of the $Na^+$ ions; or $RE^{3+}$ for three of the same. Morgan et al. [71] have utilized this phase to contain considerable amounts of both Cs and Sr and it has been evaluated by Vance and Adl [72].

## "NZP" (or sodium zirconium phosphate)

This structure appears to have no analogue in nature. It was "rediscovered" by Goodenough et al. [73] in their search for fast-ion conductors. Its structure is in some ways related to apatite but no natural mineral occurs with this structure. This is extraordinary in the light of the fact that an enormous variety of compositions have been made in this NZP structure. It is also prepared easily by ceramic techniques at low temperatures ($\sim 1000°C$). The structural formula and known substituent ions are shown below:

| VIII(?) | VI | IV |
|---------|-----|-----|
| Na | $Zr_2$ | $P_3O_{12}$ |
| Li | Ti | Si |
| K | Sn | |
| Cs | $Y^{3+}$ | |
| Ca | $Cr^{3+}$ | |
| Sr | $Al^{3+}$ | |
| | $Fe^{3+}$ | |

Moreover, it has a major advantage structurally over hollandite and magnetoplumbite in that it accommodates three size ranges of ions in these differently coordinates sites.

Alamo, Vance and Roy [74] have prepared the virtually pure $CsZr_2P_3O_{12}$ in ceramic form. Mat Kari et al. [75] have prepared single crystals of the phase. Various Sr-containing compositions and phases containing both Cs and Sr have also been made by Alamo et al. As a single ceramic phase it is unique among candidate radiophases prepared so far in its ability to accomodate a wide variety of radionuclides.

REFERENCES

1. S. S. Goldich, "A Study of Rock Weathering." J. Geol. 46:17-58 (1938).

2. F. C. Loughnan, Chemical Weathering of the Silicate Minerals. American Elsevier, New York (1969).

3. M. L. Jackson, S. A. Tyler, A. L. Willis, G. A. Bourbeau, and R. P. Pennington, "Weathering Sequence of Clay-Size Minerals in Soils and Sediments. I. Fundamental Generalizations." J. Phys. Coll. Chem. 52:1237-1260 (1948).

4. M. L. Jackson, Y. Hseung, R. B. Corey, E. J. Evans, and R. C. Henral, "Weathering Sequence of Clay Size Minerals in Soils and Sediments. II. Chemical Weathering of Layer Silicates." Proc. Soil Sci. Soc. Amer. 16:3-6 (1952).

5. M. L. Jackson and G. D. Sherman, "Chemical Weathering of Minerals in Soils." Adv. Agronomy 5:219-318 (1953).

6. F. J. Pettijohn, "Persistence of Heavy Minerals and Geologic Age." J. Geol. 49:610-625 (1941).

7. R. M. Garrels and C. L. Christ, Solutions, Minerals and Equilibria. Harper and Row, New York (1965).

8. D. Rai and W. L. Lindsay, "A Thermodynamic Model for Predicting the Formation, Stability and Weathering of Common Soil Minerals." Soil Sci. Soc. Amer. Proc. 39:991-996 (1975).

9. A. E. Nielsen, Kinetics of Precipitation. Macmillan, New York (1964).

10. A. W. Hofmann, B. J. Giletti, H. S. Yoder, Jr., and R. A. Yund, Geochemical Transport and Kinetics. Carnegie Institution of Washington, Pub. 634, Washington, D.C. (1974).

11. R. D. Shannon and C. T. Prewitt, "Effective Ionic Radii in Oxides and Fluorides." Acta Cryst. B25:925-946 (1969).

12. W. A. Deer, R. A. Howie and J. Zussman, Rock-Forming Minerals. John Wiley, New York (5 Volumes) (1962).

13. J. D. Scott, "The Crystal Structure of Miserite, A Zoltai Type 5 Structure." Canadian Mineral. 14:515-528 (1976).

14. Sridhar Komarneni, Gregory J. McCarthy and Sarah Ann Gallagher, "Cation Exchange Behavior of Synthetic Cesium Aluminosilicates." Inorg. Nucl. Chem. Letters 14:173-177 (1978).

15. D. W. Breck·and J. V. Smith, "Molecular Sieves." Sci. Amer. 44:85-91 (1959).

16. D. H. Olsen, G. T. Kokotailo and J. F. Charnell, "Crystal Structure of Cerium (III) Exchanged Faujasite." Nature 215:270-271 (1967).

17. K. A. Vlasov, Geochemistry and Mineralogy of Rare Elements and Genetic Types of Their Deposits. Vol. 2, Mineralogy of the Rare Elements. Monson Wiener Bindery, Jerusalem (1966).

18. R. C. Ewing, "Alteration of Metamict Rare-Earth $AB_2O_6$ Type Na-Ta-Ti Oxides." Geochim. Cosmochim. Acta 39:521-530 (1975).

19. L. V. Wambeke, "The Alteration Processes of the Complex Titano-Niobo-Tantalates and Their Consequences." Neues Jahr. Mineral. Abh. 112:117-149 (1970).

20. B. H. Flinter, J. R. Butler and G. M. Harrall, "A Study of Alluvial Monazite from Malaya." Amer. Mineral. 48:1210-1226 (1963).

21. D. Langmuir, "Uranium Solution-Mineral Equilibria at Low Temperatures with Applications to Sedimentary ore Deposits." Geochim. Cosmochim. Acta 42:547-569 (1978).

22. F. Smithson, "The Alteration of Detrital Minerals in the Mesozoic Rocks of Yorkshire." Geol. Mag. 78:97-112 (1941).

23. E. R. Graham, "The Plagioclase Feldspars as an Index to Soil Weathering." Soil Sci. Soc. Amer. Proc. 14:300-302 (1950).

24. G. B. Naumov, B. V. Ryzhenko and I. L. Khodakovsky, Handbook of Thermodynamic Data. National Technical Information Service, U.S. Department of Commerce (1974).

25. S. H. Chien, "Thermodynamic Considerations on the Solubility of Phosphate Rock." Soil Science 123:117-121 (1977).

26. L. Dryden and C. Dryden, "Comparative Rates of Weathering of Some Common Heavy Minerals." J. Sedimentary Petrol. 16:91-96 (1946).

27. O. Muller, W. B. White and R. Roy, "Crystal Chemistry of Some Technetium-Containing Oxides." J. Inorg. Nucl. Chem. 26:2075-2086 (1964).

28. C. Palache, H. Berman and C. Frondel, The System of Mineralogy, Vols. I & II. John Wiley, New York (1944, 1951).

29. H. Strunz, Mineralogische Tabellen. Adademisch Verlagsgesellschaft Geestu. Pontig, Liepzig, Germany (1970).

30. W. L. Robert, G. R. Rapp, Jr., and J. Weber, Encyclopedia of Minerals. Van Nostrand Reinhold Co., New York (1974).

31. W. C. Broegger, "Amorf," Salmonsens Store Illustrerede Konversationslexikon 1:742-743 (1893).

32. J. Graham and M. R. Thorber, "The Crystal Chemistry of Complex Niobium and Tantalum Oxides, IV. The Metamict State." Am. Mineral. 59:1047-1050 (1974).

33. R. C. Ewing, "The Crystal Chemistry of Complex Niobium and Tantalum Oxides, IV. The Metamict State: Discussion." Am. Mineral. 60:728-733 (1975).

34. A. Pabst, "The Metamict State." Am. Mineral. 37:137-157 (1952).

35. J. Berman, "Identification of Metamict Minerals by X-ray Diffraction." Am. Mineral. 40:802-827 (1955).

36. J. Lima-de-Faria, "Identification of Metamict Minerals by X-ray Powder Photographs." Junta de Investigacoes do Ultramar, Estudos, Ensaios e Documentos, Lisbon, No. 112, 74 pp. (1964).

37. R. S. Mitchell, "Virginia Metamict Minerals: Euxenite and Priorite." Southeastern Geology 14:59-72 (1972).

38. Tateo Ueda and Masaaki Korekawa, "Metamictization." Memoirs of College Science, University of Kyoto, Series B, Vol. 21, pp. 151-161 (1954).

39. P. F. Kerr and H. D. Holland, "Differential Thermal Analysis of Davidite." Am. Mineral. 36:563-572 (1951).

40. J. Orcel, "Analyse thermique différéntiell de quelques mineraux métamictes." Acad. Sci. Paris, Compt. Rend. 236:1052-1054 (1953).

41. S. F. Kurath, "Storage of Energy in Metamict Minerals." Am. Mineral. 42:91-99 (1957).

42. O. Mügge, "Über isotrop gewordene Kristalle." Akad. Wiss. Gottingen Math.-Phys. Kl. Nachr. 2:110-120 (1922).

43. W. Primak, "The Metamict State." Phys. Rev. 95:837 (1954).

44. Vladimiř Bouška, "A Systematic Review of Metamict Minerals." Acta Universitatis Carolinae-Geologica 3:143-169 (1970).

45. R. C. Ewing, "Metamict Columbite Re-examined." Mineral. Mag. 40:898-899 (1976).

46. H. D. Holland and D. Gottfried, "The Effect of Nuclear Radiation on the Structure of Zircon." Acta Cryst. 8:291-300 (1955).

47. A. A. Krasnobayev, Yu. M. Polezhayev, B. A. Yunikov and B. K. Novoselov, "Laboratory Evidence on Radiation and the Genetic Nature of Metamict Zircon." Geochem. Int. 11:195-209 (1974).

48. C. E. B. Conybeare and R. B. Ferguson, "Metamict Pitchblende from Goldfields, Saskatchewan and Observations on Some Ignited Pitchblendes." Am. Mineral. 35:401-406 (19  ).

49. E. J. Brooker and E. W. Nuffield, "Studies of Radioactive Compounds: IV. Pitchblende from Lake Athabasca, Canada." Am. Mineral. 37:363-385 (1950).

50. C. O. Hutton, "Manganomossite Restudied." Am. Mineral. 44:9-18 (1959).

51. J. B. Higgins and P. H. Ribbe, "The Crystal Chemistry and Space Groups of Natural and Synthetic Titanates." Am. Mineral. 61:878-888 (1976).

52. A. Faessler, "Untersuchungen zem Problem des metamikten Zustandes." Zeit. für Krystallographie 104:81-113 (1942).

53. G. A. Sidorenko, "X-ray Diffraction Identification of Metamict Minerals." Rentg. Mineral. Syrya 3:55-65 (1963).

54. D. R. Peacor and W. B. Simmons, Jr., "New Data on Graphite." Am. Mineral. 57:269-272 (1972).

55. R. C. Ewing, "A Numerical Approach Toward the Classification of Complex, Orthorhombic, Rare-Earth $AB_2O_6$-Type Na-Ta-Ti Oxides." Canadian Mineralogist 14:111-119 (1976).

56. E. I. Nefedov, "New Data of Fergusonite and Euxenite." Inform. Sbornik, Vses. Nauchn.-Issled. Geol. Inst. 3:82-85 (1956).

57. V. B. Alexandrov, "The Crystal Structure of Fersmite." Acad. Nauk SSR Doklady 132:669-672 (19  ).

58. V. B. Alexandrov, "The Crystal Structure of Aeschynite." Akad. Nauk SSR Doklady 149: 181-184 (1962).

59. A. G. Zhabin, V. B. Alexandrov, M. E. Karakovo and V. B. Feklichev, "Aeschynite from Arfvedsonite-Calcite-Quartz Veins, Visnevye Mts., Urals." Akad. Nauk SSSR Doklady 143: 686-689 (1966).

60. František Čech, S. Vrána and P. Povondra, "A Non-metamict Allanite from Zambia. N. J. Mineral. 116:208-223 (1972).

61. S. A. Gorzhevskaya and G. A. Sidorenko, "A Crystalline Variety of Lyndochite." Acad. Nauk SSSR Doklady 146:1176-1178 (1962).

62. T. Ueda, "Studies of Metamictization of Radioactive Minerals." Memoirs of College of Science, University of Kyoto, Series B, Vol. 24 (2):81-120 (1957).

63. L. M. Lipova, "Mechanism of Metamict Mineral Alteration." Geokhimiya 6:729-733 (1966).

64. P. M. Hurley and H. W. Fairbain, "Radiation Damage in Zircon, A Possible Age Method." Bull. Geol. Soc. Am. 64:659-673 (1953).

65. C. O. Hutton, "Heavy Detrital Minerals." Geol. Soc. Am. Bull. 61:635-710 (1950).

66. R. C. Ewing, "Spherulitic Recrystallization of Metamict Polycrase." Science 184:561-562 (1974).

67. T. E. Krogh and G. L. Davis, "Alteration in Zircons and Differential Dissolution of Altered and Metamict Zircon." Annual Report of the Director Geophysical Laboratory, Carnegie Institution, pp. 619-623 (1975).

68. E. S. Larsen, C. L. Waring and J. Berman, "Zoned Zircon from Oklahoma." Am. Mineral. 38:1118-1125 (1953).

69. F. A. Mumpton and R. Roy, "Hydrothermal Stability Studies of the Zircon-Thorite Group." Geochim. Cosmochim. Acta 21:217-238 (1961).

70. A.E. Ringwood (1978), Safe Disposal of High Level Nuclear Reactor Wastes: A New Strategy, A.N.U. Press, Canberra, Australia.

71. P.E.D. Morgan, D.R. Clarke, C.M. Jantzen and A.B. Harker, "High-Alumina Tailored Nuclear Waste Ceramics," J. Amer. Ceram. Soc. 64:240 (1981).

72. E.R. Vance and T. Adl (1982), Leaching Studies of Crystalline Sodium Phases and Nuclear Waste Forms, Scientific Basis for Nuclear Waste Management, Vol. 4, Ed. C.J. Northrup, Jr., Plenum Press, NY.

73. J.B. Goodenough, H.Y.P. Hong and F.A. Kafalas, "Fast $Na^+$-ion Transport in Skeleton Structures, Mat. Res. Bull. 11:203 (1976).

74. J. Alamo-Serrano, E.R. Vance, and R. Roy, "A New Versatile Ceramic Radiophase Structure, NZP," Mat. Res. Bull. (in preparation).

75. B. Mat Kari, B. Prodi and M. Sljakus, "Preparation and Structural Studies of Phosphates with Common Formula $M^I M_2^{IV}(PO_4)_3$ ($M^I$ = Li,Na,K,Rb,Cs;$M^{IV}$ = Th,U,Zr,Hf), Bull. Soc. Chem. France, 1777 (1978).